John A. S. Ross
January '82
Harrow

GW00722758

Receptors and Recognition

General Editors: P. Cuatrecasas and M.F. Greaves

About the series

Cellular Recognition – the process by which cells interact with, and respond to, molecular signals in their environment – plays a crucial role in virtually all important biological functions. These encompass fertilization, infectious interactions, embryonic development, the activity of the nervous system, the regulation of growth and metabolism by hormones and the immune response to foreign antigens. Although our knowledge of these systems has grown rapidly in recent years, it is clear that a full understanding of cellular recognition phenomena will require an integrated and multidisciplinary approach.

This series aims to expedite such an understanding by bringing together accounts by leading researchers of all biochemical, cellular and evolutionary aspects of recognition systems. This series will contain volumes of two types. First, there will be volumes containing about five reviews from different areas of the general subject written at a level suitable for all biologically oriented scientists (Receptors and Recognition, series A). Secondly, there will be more specialized volumes (Receptors and Recognition, series B), each of which will be devoted to just one particularly important area.

Advisory Editorial Board

Receptors and Recognition

Series A

Published

Receptors and
Recognition

Series B Volume 13

Receptor Regulation

Edited by
R. J. Lefkowitz

Professor of Medicine
Duke University Medical Center, Durham
North Carolina

LONDON NEW YORK

CHAPMAN AND HALL

First published 1981
by Chapman and Hall Ltd,
11 New Fetter Lane, London EC4P 4EE

Published in the U.S.A. by
Chapman and Hall
in association with Methuen, Inc.,
733 Third Avenue, New York, NY 10017

Typeset by Preface Ltd, Salisbury, Wilts.
and printed in Great Britain
at the University Printing House, Cambridge

ISBN 0 412 15930 9

British Library Cataloguing in Publication Data

Receptor regulation – (Receptors and
recognition.
   Series B; Vol. 13)
   1. Cell receptors
   I. Lefkowitz, Robert J. II. Series
   574.8'75   QH603.C43

   ISBN 0-412-15930-9

Contents

Contributors

G. Carpenter, Department of Biochemistry, Vanderbilt University School of Medicine, Nashville, Tennessee, U.S.A.

S. Cohen, Department of Medicine (Dermatology), Vanderbilt University School of Medicine, Nashville, Tennessee, U.S.A.

A.O. Davies, Department of Medicine, Duke University Medical Center, Durham, North Carolina, U.S.A.

A. Engel, Mayo Clinic, Rochester, Minnesota, U.S.A.

D.M. Fambrough, Department of Embryology, Carnegie Institution of Washington, Baltimore, Maryland, U.S.A.

H.G. Friesen, Department of Physiology, University of Manitoba, Winnipeg, Manitoba, Canada.

R.J. Lefkowitz, Department of Medicine, Duke University Medical Center, Durham, North Carolina, U.S.A.

J. Lindstrom, Salk Institute for Biological Studies, San Diego, California, U.S.A.

F. de Pablo, National Institute of Arthritis, Metabolism and Digestive Diseases, National Institutes of Health, Bethesda, Maryland, U.S.A.

B. Rees Smith, Department of Medicine, Welsh National School of Medicine, University Hospital of Wales, Cardiff, U.K.

J. Roth, National Institute of Arthritis, Metabolism and Digestive Diseases, National Institutes of Health, Bethesda, Maryland, U.S.A.

R.P.C. Shiu, Department of Physiology, University of Manitoba, Winnipeg, Manitoba, Canada.

E. Van Obberghen, National Institute of Arthritis, Metabolism and Digestive Diseases, National Institutes of Health, Bethesda, Maryland, U.S.A.

1 Introduction

ROBERT J. LEFKOWITZ, M.D.

Receptor Regulation
(*Receptors and Recognition*, Series B, Volume 13)
Edited by R. J. Lefkowitz
Published in 1981 by Chapman and Hall, 11 New Fetter Lane, London EC4P 4EE
© 1981 Chapman and Hall

The study of hormone and drug receptors has become one of the most exciting and rapidly moving areas of biomedical research. Elucidation of receptor mechanisms and receptor structure has become the common goal of many scientists from diverse backgrounds. The rapid advances achieved have been due, in large part, to the concentrated effort of workers from a variety of disciplines including classical pharmacology, biochemistry, endocrinology, cell biology, genetics and molecular biology, among others.

Hormone and drug receptors appear to be of three major types, which may be classified by their cellular locations. Found in the plasma membranes of cells are the receptors for a wide variety of polypeptide hormones, catecholamines and a variety of neurotransmitters. Included within this group are those receptors coupled to the enzyme adenylate cyclase. The second group of receptors are the soluble cytoplasmic receptors for the steroid hormones. A third type of hormone receptor is the receptor for the thyroid hormones which appears to be confined to the nucleus. Not only may these different types of receptor be distinguished in terms of their cellular locations but also by their mechanisms of action. Thus, the plasma membrane receptors appear to function by triggering the release of second messengers such as cyclic AMP or calcium which then regulate a variety of cellular metabolic events. The soluble steroid receptors bind steroid hormones in the cytoplasm, become activated and then enter the nucleus where they alter transcription of the genome. The thyroid hormones bind directly to nuclear receptors and also seem to alter transcription.

The modern era of receptor research can be envisaged as beginning with the advent of successful direct radioligand binding studies which permitted a direct approach to the investigation of the receptors. Well-validated radioligand binding assays for the study of almost all the known receptors are currently available. A wide variety of high affinity, high specific radioactivity radioligands are available for the investigation of these receptors.

One of the most significant insights to develop from ligand binding studies of hormone and drug receptors is that the receptors, rather than representing static 'receiving stations' for hormonal signals, in fact are subject to very dynamic regulation. This dynamic regulation of receptor number and properties likely represents one of the most important mechanisms which have evolved for the control of tissue sensitivity to humoral and neurohumoral messages. The present volume attempts to distill the essence of recent research concerning 'receptor regulation'. Several words about its organization are in order. First, this field is already so vast that an encyclopaedic approach is not feasible. Accordingly, representative areas of research have been reviewed which seem best to highlight the important principles which

3

are emerging. The first section of the book contains three chapters dealing with receptors for the peptide 'hormones' insulin, epidermal growth factor and prolactin. The second section contains two chapters dealing with receptors for the neurotransmitters, catecholamines and acetylcholine. The essays in these first two sections serve to underscore the complexity and multiplicity of mechanisms which may regulate receptor properties. Several interesting and important principles emerge. Hormones are found generally to regulate their own receptors. As first shown for the insulin receptors by Roth's group there seems to be an inverse correlation between the ambient level of hormone or neurotransmitter and the density of receptors on cells. Thus, the higher the agonist concentration the lower the receptor concentration. This appears to function as a feedback mechanism to dampen cellular sensitivity to hormonal or drug action in the face of high concentrations of biologically active agents. In the case of the acetylcholine receptor it is found that chronic nerve stimulation also leads to a compensatory decrease in receptor number. Mechanisms by which hormones 'down regulate' their receptors seem complicated and as yet are not fully delineated. These apparently involve hormone promoted internalization of receptors and changes in the rates of receptor synthesis, as well as changes in the rates of receptor degradation. The receptors for prolactin and angiotensin appear to be exceptions, since these receptors are 'up-regulated' by the natural hormone. In some cases antagonists may lead to an opposite change in receptor number; thus high concentrations of antagonists up-regulate receptors, possibly leading to increased sensitivity to hormone or drug effect.

In addition to regulation by 'homologous' hormones, receptors appear also to be regulated by a wide variety of hormones with which they do not normally combine. For example, thyroid hormone seems to control receptors for the catecholamines and prolactin among others. Steroid hormones are found to control the concentrations and properties of a variety of receptors for hormones and neurotransmitters which are found in the plasma membranes. Such 'heterologous' regulation is apparently a very common pattern for regulating the number of receptors.

In addition, a wide variety of other influences appear to control receptor properties. These include the cell cycle, mitogenesis and the state of differentiation or dedifferentiation of a tumor to name just a few. Many other examples of interesting receptor regulatory phenomena are described and discussed in detail in these chapters.

In the final section of the book are found three chapters dealing with clinical situations in which receptor alterations are fundamentally involved in the pathophysiological basis of several diseases. The theme which runs through these chapters is that anti-receptor auto-antibodies can lead in a variety of interesting ways to manifestations of disease. Thus, in the case of the insulin receptor and the nicotinic cholinergic receptor, antireceptor antibodies interfere with the normal functioning of the receptor and lead respectively to either a diabetic state or to myasthenia gravis. An interesting contrast is provided by the anti-TSH receptor antibody found in the serum of patients with Graves Disease, which actually stimulates the receptor and leads to increased production of thyroid hormone.

Another fascinating aspect of this area of research, which is particularly illuminated by the studies of the insulin receptor and the nicotinic cholinergic receptor, is how antireceptor antibodies can be used as powerful investigative tools to probe the structure and function of hormone and drug receptors.

Viewed together, the individual chapters of this book provide a picture of the recently discovered mechanisms which regulate receptors for hormones and neurotransmitters. They demonstrate how an important level of control of cellular metabolism and function (the receptors) may be rapidly modulated in a variety of normal and pathophysiological states.

2 The Insulin Receptor and its Function

EMMANUEL VAN OBBERGHEN
and JESSE ROTH

Receptor Regulation
(*Receptors and Recognition*, Series B, Volume 13)
Edited by R. J. Lefkowitz
Published in 1981 by Chapman and Hall, 11 New Fetter Lane, London EC4P 4EE
© 1981 Chapman and Hall

2.1 INSULIN ACTION

The sequence of events for insulin action on its target cells is schematically represented in Fig. 2.1. After being secreted by the pancreatic β cells, insulin travels in the plasma to target cells throughout the body, where it binds rapidly and reversibly to a finite number of receptors on the external surface of the plasma membrane. The receptor serves two functions. First, the receptor recognizes insulin from among all the other substances to which the cell is being exposed; it manifests its recognition by binding to the hormone. Second, the hormone–receptor complex begins the activation at the target cell that leads ultimately to the multiplicity of effects that insulin produces on any of its target cells. It is widely believed that insulin action on its target cells is mediated by a soluble intracellular messenger that functions to carry out all the effects of insulin. While it is clear that cyclic AMP is not the second messenger for insulin, it is suspected that the second messenger for insulin functions in a manner analogous to that described for cyclic AMP (Roth, 1979a). Recent studies by two laboratories have provided evidence for the existence of a second messenger for insulin, a soluble product thought to be produced from endogenous components of the target cell that carries out fundamental actions of insulin, such as enzyme activation and inactivation (Seal and Jarrett, 1980; Larner *et al.*, 1980). Whereas hormones

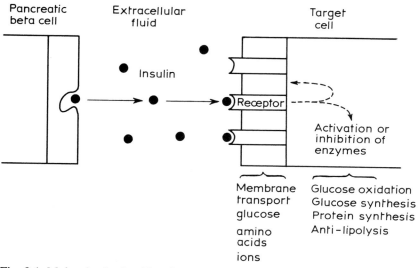

Fig. 2.1 Molecular basis of insulin action.

9

that work through cyclic AMP stimulate phosphorylation of target cell components and thereby activate and inactivate enzymes, insulin action generally does the opposite, i.e. stimulates dephosphorylation and thereby activates and inactivates enzymes in a fashion opposite to that induced by cyclic AMP-dependent protein kinases.

2.2 FUNCTIONS OF THE RECEPTOR

2.2.1 Information transfer

In a biological sense it is clear that both the hormone and the receptor are required to activate the target cell. However, it seems useful to try to distinguish whether the information for activation is contained within the hormone or in the receptor, or whether both components contribute about equally. As can be seen in Table 2.1, in some communication systems both ligands contribute equally. In other systems it is clear that the ligand has the message and that the receptor serves largely to concentrate the ligand, to help process it and transfer it to internal or intracellular sites where the ligand then acts. Examples where the ligand has all the information include toxins, viruses, lysosomal enzymes and cholesterol-containing plasma proteins. Initially it had been thought that insulin might act in an analogous fashion. In fact, reports appeared suggesting that some components of the insulin molecule were generated which mediated insulin action. However, recent studies have failed to substantiate these early reports. Furthermore,

Table 2.1 Information transfer

A. Hormone + Receptor $\rightleftarrows$ Hormone–Receptor complex $\rightarrow$ activation
B. Where is the *'information'* for the activation – H?? R?? HR??
C. Examples of ligand–receptor systems
    1. Both moieties contribute about equally – egg and sperm
    2. Information is in the ligand; receptor acts to concentrate, process, and/or translocate the ligand to an intracellular site
       (a) toxins: cholera, diphtheria, ricin
       (b) low-density lipoproteins
       (c) asialo-glycoproteins
    3. Information is in the receptor; ligand acts to bring out the program from the receptor
       (a) insulin
       (b) IgE
       (c) (thyrotropin)
       (d) (acetylcholine)

Table 2.2 Functions of the receptor–hormone interaction

A. Fundamental
   1. Recognition
   2. Activation
B. Additional
   1. Reservoir for plasma hormone
   2. Regulate degradation of hormone
   3. Regulate degradation of receptor
   4. Regulate receptor concentration and receptor affinity
   5. Cross-link and translocate hormone–receptor complexes
   6. Regulate post-receptor events

---

recent studies have shown that insulin molecules that have been synthesized lacking this putative active component retain the biological activity of insulin.

At present, we think that the receptor for insulin contains the full program to activate the cell (Roth, 1979b). This is based on studies of anti-receptor antibodies. These anti-receptor antibodies, produced spontaneously in patients with auto-immune disorders or produced in experimental animals by immunization with receptor preparations, bind to the receptor and produce activation of all the biochemical pathways typical of insulin action on that target cell. This includes not only all events at the cell surface such as glucose transport and amino acid transport, but also activation of glycogen synthase and pyruvate dehydrogenase as well as delayed effects, such as stimulation of the synthesis of lipoprotein lipase. The observation that anti-receptor antibodies in the absence of insulin are capable of generating specifically the full program of insulin action suggests to us that the receptor has all the information and that insulin serves largely or exclusively in binding to the receptor to bring out that program from the receptor. Lectins, which bind to the receptor, are also capable of generating insulin-like programs.

2.2.2 Other functions of the hormone–receptor complex

The interaction of insulin with its receptor serves not only for purposes of recognition and activation, but also acts to regulate the affinity of the insulin receptor (by negatively cooperative site–site interactions (De Meyts *et al.*, 1973)), to regulate the concentration of the receptor (Gavin *et al.*, 1974), and to modulate target cell sensitivity in a variety of ways at multiple post-receptor sites. In addition, the binding of the hormone to the cell surface receptor acts as a reservoir for plasma hormone, binding hormone as the hormone concentration rises and releasing hormone back into the plasma as hormone concentration falls (Zeleznik and Roth, 1978). Finally, a major pathway for the degradation of insulin *in vivo* appears to be carried out *via* a receptor-mediated pathway (Terris and Steiner, 1975), (Table 2.2).

2.2.3 Other features of the hormone–receptor interaction

Insulin receptors are found not only on the plasma membrane of the cell, but also on other membranous structures, including rough endoplasmic reticulum, Golgi, and, possibly, on the nuclear envelope (Bergeron *et al.*, 1973; Kahn *et al.*, 1973; Horvat and Katsoyannis, 1975; Goldfine and Smith, 1976; Kahn, 1976). The source, fate and function of these intracellular receptors is under intense study but many of the fundamental issues remain unresolved. Do they represent newly synthesized receptors on the way to the plasma membrane? Do they represent receptors that had been on the cell surface that are in the process of being internalized or on their way to be degraded or on the way back to be reinserted in the plasma membrane? While it is clear that receptors exist on intracellular structures and that hormone can be recovered from intracellular sites, it is generally agreed that the fundamental interaction of hormone with receptor occurs initially at the cell surface (Kahn, 1976; Roth, 1979a).

While the initial interaction of insulin with its receptor on the cell surface is a reversible interaction, it is clear that at 37° C multiple other events occur quickly. For example, there may be irreversible binding of hormone to receptor. Hormone–receptor complexes may aggregate and within a short time some of these complexes may appear at intracellular loci (Gorden *et al.*, 1978; Schlessinger *et al.*, 1978, 1980; Carpentier *et al.*, 1978; Barazzone *et al.*, 1980). In some studies the intracellular hormone appears to be localized to intracellular regions near the cell surface associated with lysosome-like vesicles (Carpentier *et al.*, 1979). Other investigators have suggested that these vesicular structures may represent vesicles that are derived from Golgi components (Bergeron *et al.*, 1979). Some investigators have found insulin associated with the nuclear envelope (Goldfine *et al.*, 1978), although others have claimed that this may represent an artifact of tissue preparation. The mechanism by which surface-associated insulin is internalized has been studied extensively and it has been suggested that the internalization process takes place in the region of coated pits (Suzuki and Kono, 1979; Goldstein *et al.*, 1979; Schlessinger, 1980; Carpentier *et al.*, 1980). There seems to be general agreement that the hormone internalized by the receptor mechanism is degraded relatively rapidly at intracellular sites. Whether the receptor that is internalized is recirculated back to the cell surface or is itself also degraded is as yet unclear. It is also unclear which of these translocations of the hormone and of the receptor are needed for the various functions of the hormone–receptor complex. While the issue is still not yet completely settled, it appears that the initial interaction of insulin with its receptor at the cell surface generates the fundamental signal for activation of the cell and that the other processes, such as aggregation and internalization, act to modify the activation process or to provide for regulation of hormone and receptor, but are not needed for the fundamental activation of the target cell.

2.3 CHARACTERISTICS OF THE INSULIN RECEPTOR

The insulin receptor has not yet been purified sufficiently to allow direct chemical characterization. Thus, most of our information about its structure is indirect (Kahn, 1976; Roth, 1979a). The receptor is considered to be an 'integral' membrane glycoprotein, since it cannot be removed from the membrane except by use of detergents. Presumably, the receptor has an hydrophobic region buried in the membrane. The receptor contains proteins, as evidenced by its susceptibility to destruction by proteolytic enzymes (Kono, 1971; Kono and Barham, 1971). The receptor presumably also contains carbohydrate components since it binds to various lectins (Cuatrecasas and Tell, 1974; Cuatrecasas, 1974). However, there is no evidence indicating that these carbohydrates participate directly in the insulin binding site.

When studied in the presence of neutral detergents, insulin binding is essentially unchanged and the insulin receptor behaves as a large molecule with a Stokes radius of 68–72 Å and a molecular weight of 300 000 to 1 000 000 (Gavin *et al.*, 1971; Cuatrecasas, 1972; Ginsberg *et al.*, 1976; Maturo and Hollenberg, 1978; Lang *et al.*, 1979). Upon the addition of high insulin concentrations, this large molecule dissociates and generates a subunit with an estimated Stokes radius of 38–40 Å, consistent with a molecular weight of 75 000 (Ginsberg *et al.*, 1976). The insulin binding activity is preserved in this smaller species. These observations are consistent with the concept that the solubilized receptor is a tetramer or exists in close association with membrane components participating in the regulation of the receptors and/or insulin action. Studies by Maturo and Hollenberg (1978) on solubilized receptors and by Harmon *et al.* (1980) on membrane-bound receptors support the existence of such regulatory components associated with the insulin receptor.

Since the insulin receptor loses its binding activity in the presence of sodium dodecyl sulphate (SDS), attempts to characterize its subunit structure have been seriously hampered. Recent observations suggest that three or four subunits form the insulin receptor. The size of the insulin binding component has been estimated to be 90 000 by radiation inactivation (Harmon *et al.*, 1980), 125 000–140 000 by binding of photoreactive insulins (Yip *et al.*, 1978, 1980; Wisher *et al.*, 1980) or by chemical cross-linking of ^{125}I-labeled insulin (Pilch and Czech, 1980; Kasuga *et al.*, 1981).

The concentration of insulin receptors ranges from 10^3 to 10^5 per cell. In contrast, the number of insulin receptors per unit cell surface is remarkably consistent for cells of varying species and tissues (Ginsberg, 1977). Most cells have 15–60 sites per square micron of membrane surface when the cells are treated as simple geometrical forms. Highly purified plasma membrane preparations also bind remarkably similar amounts of insulin (Ginsberg, 1977). The total amount of cell surface insulin receptors represents less than 1 per cent of the protein of the plasma membrane (Ginsberg, 1977).

2.4 QUANTITATIVE ASPECTS OF THE HORMONE–RECEPTOR INTERACTION

Theoretically, one can consider the interaction of a hormone with its receptor according to the scheme in Table 2.3. Free hormone (H) binds reversibly to receptors (R) to form hormone–receptor complexes (HR). The affinity of this interaction is expressed as K. The biological activity of the hormone is some function (f) of the number of HR complexes formed, which, in turn, is dependent on the hormone concentration [H], receptor concentration [R], and the affinity of the receptor for the hormone, K. The magnitude and the rate of the signal to the cell depend on these three factors, i.e. [H], [R] and K. Consequently, a change in any of these three determinants will affect the signal to the target cell.

It has been known for a long time that hormone concentrations can fluctuate rapidly *in vivo*. However, it has been recognized only recently that both the affinity and concentration of insulin receptors change in response to physiological stimuli and that these fluctuations are very important in determining the magnitude of the biological effect. Examples of biologically relevant regulators of insulin receptors are summarized in Table 2.3 and will be discussed in detail later.

2.5 METHODS OF MEASUREMENT

2.5.1 Receptor assay

The first attempts designed to measure hormone receptors were reported nearly three decades ago, but all of them were unsuccessful (Stadie *et al.*, 1953; Newerly and Berson, 1957), for several reasons: (i) poor biological activity of the labeled hormone; (ii) degradation of both hormone and receptor, and (iii) high non-specific binding, which made it impossible to decide whether the hormone binding to target tissues was biologically relevant. In the 1960s studies were specifically designed to eliminate these

Table 2.3 The interaction of hormone with receptor*

1. The primary event:

$$H + R \leftrightharpoons HR$$

2. The equilibrium constant (K_a) for this reaction:

$$K_a = \frac{[HR]}{[H][R]}$$

3. The biological effect (E) is some function (f) of the strength of the signal to the cell; the signal strength is directly related to [HR].

$$\begin{aligned} E &= f([HR]) \\ &= f(K[H][R]) \\ &= f\frac{K[H][R_0]}{1 + K[H]} \end{aligned}$$

Thus the signal strength to the cell depends equally on K, H and R (or R_0).

* H = free hormone, R = free receptor, HR = hormone–receptor complexes, R_0 = total receptors (R + HR).

problems. Moreover, the methods presently used to measure cell-surface receptors are based on the methods introduced in the 1960s for the study of ACTH and angiotensin binding to their receptors (Lefkowitz *et al.*. 1969. 1970; Goodfriend and Lin, 1970; Lin and Goodfriend, 1970). The success of these pioneering studies resulted from (i) the preparation of labeled hormone with high specific activity and preserved biological activity; (ii) the meticulous handling of the receptor preparations, assuring low degradation of hormone and receptor and low non-specific binding, and (iii) the demonstration that the hormone specificity of the binding matched that of the biological system, i.e. bioactive analogs competed for binding in direct proportion to their biological potency, and unrelated hormones or irrelevant materials did not compete. The techniques originally developed for ACTH and angiotensin were later applied with success to a variety of peptide hormones, catecholamines, neurotransmitters, prostaglandins, lectins, toxins, lipoproteins, microbes and other agents with cell-surface receptors.

The methodology of the radioreceptor assay is similar to the classic radioimmunoassay technique or other competition binding assays except that a source of receptors (e.g. isolated whole cells or plasma membranes) is used in place of the hormone antibody or binding protein (Roth, 1973, 1979a; De Meyts, 1976). Freshly isolated or cultured cells, or membrane-enriched fractions obtained from ruptured cells, or highly purified membrane preparations, or detergent-solubilized cells or membranes are used as receptor source. Typically, a small 'tracer' amount of ^{125}I-labeled hormone and increasing amounts of unlabeled hormone are allowed to compete for receptor binding under steady-state conditions, standardized as regards temperature, pH, ionic strength, etc. The conditions are also chosen to minimize degradation of both the receptor and the hormone. The receptor-bound hormone and the free hormone are then separated and the radioactivity associated with the receptors is counted. The technique applied to separate the receptor-bound hormone from the free hormone depends on the nature of the receptor. Thus, centrifugation is used in the case of particulate receptors (i.e. receptors on cells or membranes), whereas gel filtration or precipitation with polyethylene glycol* is applied in the case of solubilized receptors.

The major characteristics of the functional definition of the hormone receptor are the specificity of the binding site for the hormone and the fact that the binding of the hormone to the sites can be related to the biological effects of the hormone. This property has been extensively studied by Freychet *et al.* (1971, 1975) and Gliemann and Gammeltoft (1974), who showed that in every case studied over 40 insulin analogs display a close correlation between affinity of the analog for the insulin receptor and its biological activity *in vitro* (Fig. 2.2). In fact, neither insulin nor its receptor has an absolute specificity. Indeed, insulin has a high affinity for its own receptor but has conserved some affinity for the receptors of structurally related hormones, i.e. the insulin-like growth factors. Conversely, these insulin-like growth factors bind with a low affinity to the insulin receptor. Both receptors are specific for insulin and insulin-related peptides (i.e. they do not bind non-insulins), but among the insulins the two receptors show different preferences.

Whatever the method used to measure receptors, specificity is the most important defining characteristic of any receptor. Further, this property unequivocally distinguishes the binding of insulin to its physiological receptor from binding to inorganic adsorbants such as talc (Ginsberg, 1977).

2.5.2 Measurement of receptors using ^{125}I-labeled anti-receptor antibodies

The antibodies against the insulin receptor used for this purpose in our laboratory have been isolated from patients with an autoimmune disorder and an unusual form of diabetes characterized by an extreme insulin resistance and acanthosis nigricans (Flier *et al.*, 1975; Kahn *et al.*, 1976; Harrison and Kahn, 1980; Jarrett *et al.*, 1976; and pp. 150–55). In brief, plasma from these patients is fractionated on DEAE-cellulose; the IgG fraction, containing the anti-receptor antibody, is labeled with [125]I. The labeled anti-receptor antibody

---

* Polyethylene glycol used at appropriate concentrations precipitates receptor-bound hormone, while the free hormone remains in solution (Desbuquois and Aurbach, 1971).

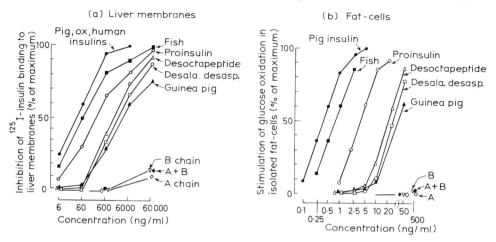

Fig. 2.2 Effect of insulin and insulin analogs on ^{125}I-insulin binding to liver membranes and glucose oxidation in fat-cells. (a) Inhibition of ^{125}I-insulin binding to rat liver membranes expressed as per cent of maximum is plotted as a function of the concentration of unlabeled peptide. (b) The stimulation of glucose oxidation by isolated rat fat-cells expressed as per cent of maximum is plotted as a function of the concentration of unlabeled peptide. Note the close correlation between affinity for the receptor (a) and biological effect (b). From Freychet *et al.* (1971).

is then selectively enriched by adsorption to cells rich in insulin receptors followed by elution from these cells. This affinity-purified anti-receptor antibody binds to a variety of cells and this binding is specifically competed for by whole plasma and purified IgG from the same patient, as well as by similar preparations from other patients with autoantibodies to the insulin receptor, but not by non-immune plasma or serum. More importantly, the ^{125}I-labeled antibody against the insulin receptor has been proven to represent a sensitive method for the detection and measurement of insulin receptors. Thus, insulin analogs that differ in their affinity for receptors, measured by inhibition of ^{125}I-insulin binding, inhibit the binding of ^{125}I-anti-receptor antibody in direct proportion to their ability to bind to the insulin receptor (Fig. 2.3). Further, there exists a close correlation between the binding of ^{125}I-insulin and ^{125}I-anti-receptor antibody over a wide range of receptor concentrations on different cell types. Finally, when the concentration of insulin receptors is reduced experimentally the ^{125}I-anti-receptor antibody binding decreases in parallel with the ^{125}I-insulin binding. In conclusion, the labeled anti-receptor antibody offers an alternate ligand for detection and quantitation of insulin receptors under circumstances where insulin binding might be altered.

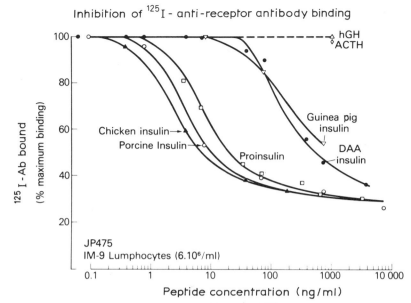

Fig. 2.3 Effect of insulin analogs on the binding of ^{125}I-anti-receptor antibody to IM-9 lymphocytes. From Jarrett *et al.* (1976).

2.5.3 Radioimmunoassay of the insulin receptor

In contrast to the preceding method, which measures the binding of ^{125}I-labeled antibody to receptor, this specific and sensitive receptor radioimmunoassay is based on the specificity of autoantibodies for immunoprecipitation of the insulin receptor. The assay is in principle quite similar to a standard radioimmunoassay except that the antigen is the solubilized insulin receptor labeled with tracer amounts of ^{125}I-insulin (Fig. 2.4) (Harrison *et al.*, 1979). Ideally, receptor that is labeled directly, e.g. with ^{125}I, would be preferable, but receptor preparations are not yet sufficiently pure for this purpose. Briefly, insulin receptors are solubilized in Triton X-100 from membranes (e.g. human placental membranes) and incubated with ^{125}I-insulin. The labeled receptor is separated from free insulin by gel filtration. A small fixed amount of the ^{125}I-insulin–receptor complex is incubated at 4° C* with anti-receptor antibody and increasing amounts of unlabeled receptor. Finally, with the addition of anti-human IgG, the ^{125}I-insulin–receptor complex bound to anti-receptor antibody is precipitated. The occurrence of significant competition for binding produced by unlabeled receptor binding

---

* The incubation is performed at low temperatures in order to minimize the dissociation of ^{125}I-insulin from the receptor.

Insulin receptor radioimmunoassay

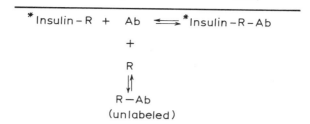

$$^*\text{Insulin}-\text{R} \;+\; \text{Ab} \;\rightleftharpoons\; ^*\text{Insulin}-\text{R}-\text{Ab}$$

Fig. 2.4 Scheme for radioimmunoassay for the insulin receptor. Receptor-antibody (Ab) is incubated with ^{125}I-insulin–receptor complex (*Insulin-R) and various concentrations of unlabeled receptor (R). Antibody-bound receptor is then precipitated by addition of anti-human IgG.

site concentrations as low as 0.1 nM attests to the sensitivity of the assay. The major advantage of this new assay is that it measures the receptor independently of its insulin binding function. The radioimmunoassay can be used to probe the structure and function of the insulin receptor. Indeed, although the binding properties of the insulin receptor are strikingly similar in all species, the radioimmunoassay uncovers immunological (i.e. structural) differences between receptors. For example, equimolar concentrations of insulin binding sites derived from different tissues (i.e. human placental membranes, cultured human IM-9 lymphocytes and mouse liver membranes) generate non-identical, but parallel displacements in the assay. Further, using the radioimmunoassay it is possible to demonstrate that the decrease in insulin binding sites occurring after chronic exposure to insulin ('down-regulation') is associated with the loss of immunoreactive receptors. The implication of this observation is that 'down-regulation' is due to the disappearance, i.e. increased degradation and/or decreased synthesis of receptor molecules, rather than to inactivation of the binding function *in situ* or to binding sites becoming inaccessible. Future useful applications of the radioimmunoassay could include the measurement of circulating or shed receptors, or inactivated receptor subunits.

2.5.4 Measurement of the whole-body insulin-receptor compartment *in vivo*

The experimental design of this technique was introduced by Zeleznik and Roth (1978), who were able to show that [125]I- or ^{131}I-insulins (i.e. pig and chicken insulin) with high affinity for receptor distribute into a body receptor compartment, which is two to three times larger than that of low affinity insulin (i.e. guinea pig insulin and proinsulin). Further, their study demonstrated that this *in vivo* body compartment displays the typical characteristics that are generally associated with specific cell surface receptors. Indeed, the whole body receptor compartment has a limited capacity to bind insulin (saturability), to interact with insulin reversibly (reversibility), and to bind insulin analogs as predicted by their biological potency (specificity).

This method was first validated by Zeleznik and Roth (1978) in rabbits, and recently extended successfully to studies in man by Wachslicht-Rodbard *et al.* (unpublished results). In brief, a labeled insulin with low affinity for receptors is intravenously injected in combination with a differently labeled insulin with high affinity for receptors. At selected time intervals following the administration of the insulin, the plasma concentrations of the insulins are measured. Based on the analysis of the time course of disappearance of each of the insulins – the total radioactivity and the intact radioactive insulin (precipitatable by trichloroacetic acid and/or anti-insulin antibodies), one can estimate the fraction of hormone bound to the specific receptors *in vivo*. The present method of measurement of insulin receptors in the whole body *in vivo* will clearly add a new and important dimension to the investigation of the hormone–receptor interactions in physiological and pathophysiological situations. The method complements the traditional method of measuring insulin receptors *in vitro* in selected tissues (Table 2.4). Its greatest value will be in humans where important target tissues for insulin (e.g. liver and muscle) are not readily available for study.

2.5.5 Localization of insulin receptors

To localize the cellular sites of hormone binding, investigators have used radioactively labeled hormones with electron-microscopic radioautography (Gorden *et al.*, 1978; Carpentier *et al.*, 1978), ferritin-labeled insulin with conventional electron microscopy (Orci *et al.*, 1975; Jarett and Smith, 1977) and hormone labeled with fluorescent dyes for localization of fluorescence (Schlessinger *et al.*, 1978).

Table 2.4 Cells from humans–insulin receptors

Fresh cells–monocytes
–erythrocytes
–adipocytes
–placenta
Cultured Cells–fibroblasts
–B-lymphocytes

2.6 BIOLOGICALLY RELEVANT FLUCTUATIONS IN RECEPTOR CONCENTRATION AND AFFINITY

Like other cellular components, insulin receptors are in a state of continual synthesis and degradation. The concentration of insulin receptors on the cell membrane reflects a state of continuing turnover, which, under steady-state conditions, is characterized by a dynamic equilibrium between the rate of receptor degradation and the rate of receptor synthesis, intracellular translocation and insertion into the cell membranes. Changes in any of these processes will lead to alterations in the concentration of insulin receptor.

A variety of biologically relevant situations is associated with changes in the concentration and affinity of the insulin receptor at steady state (Table 2.5).

2.6.1 Receptor turnover

At the present time very little information concerning the turnover of the insulin receptor is available. The decay of the insulin receptor in 3T3-L1 fatty fibroblasts (Karlsson *et al.*, 1978) and in IM-9 lymphocytes (Kosmakos and Roth, 1980) has been measured in the presence of cycloheximide and indicated a $t_{1/2}$ of 20–40 hours. Using a very elegant technique of density-labeling of receptors in 3T3-L1 adipocytes. Reed and Lane (1980) have recently determined a $t_{1/2}$ for receptor degradation of 6.7 hours. This estimation of the half-life of the insulin receptor is thus much shorter than the $t_{1/2}$ found in the presence of cycloheximide. This difference might be explained by the possibility that inhibitors of protein synthesis, such as cycloheximide, may cause a block or inhibition of the synthesis of short-lived proteins, which regulate the insulin receptor turnover (Reed and Lane, 1980). An apparent short half-life (i.e. 9 hours) of disappearance of cell surface insulin was also found when 3T3-L1 adipocytes were exposed to tunicamycin, which inhibits protein glycosylation (Rosen *et al.*, 1979). These data suggest that glycosylation may be a rate-limiting step in the provision of functional insulin binding

Table 2.5 Biologically relevant regulators of insulin receptors

1. Insulin	6. Food	7. Cell program
2. Other hormones	(a) calories	(a) cell type
3. Drugs	(b) nutrients	(b) differentiation
4. pH	(c) fiber	(c) growth
5. Exercise		(d) transformation

sites to the cell membrane. In the presence of tunicamycin the normal processes would be inhibited, resulting in the formation of inactive receptors, which do not bind the hormone.

2.6.2 Cell program

Multiple important changes in the program of the cell have been shown to result in major changes in the concentration of insulin receptors. Thus, changes in growth rate, cell cycle, cell differentiation, and culture conditions can alter the number of receptors per cell. Thomopoulos *et al.* (1976) and Bar *et al.* (1976) have demonstrated that rapidly growing cells have fewer insulin receptors than cells at stationary phase. Likewise, spontaneous-, chemical-, or virus-transformed fibroblasts which lack contact inhibition of growth, also have a lower concentration of insulin receptors when compared with non-transformed cells (Thomopoulos *et al.*, 1976). Also, the cell cycle is associated with changes in insulin receptor concentration. Thus, in fibroblasts there is an increase in receptors at the beginning of the G_1-phase, so that at stationary phase, when all the cells are at G_0, the insulin receptor concentration is maximal (Thomopoulos *et al.*, 1977). The precise mechanism responsible for the increase in insulin receptor concentration in stationary confluent cells is unknown at the present time. A possible participation of cAMP has been suggested based on the following observations. The intracellular concentration of cAMP is elevated in cells at confluence (Pastan *et al.*, 1975). Similarly, cAMP itself is capable of increasing receptor concentration twofold in stationary confluent cells. However, since the increase in receptor number observed at confluence is two- to four-fold greater than the increase found following preincubation with cAMP, Thomopoulos *et al.* (1977) concluded that the spontaneous increase in receptor number occurring at confluence can be accounted for only in part by the increased levels of cAMP. Note that the conclusion drawn by Thomopoulos *et al.* (1977) that the concentration of insulin receptor is elevated in stationary cells and low in growing and transformed cells, apparently contradicts the study by Krug *et al.* (1972). The latter reported the absence of specific binding sites for insulin on resting circulating human lymphocytes. However, after lymphoblastic transformation *in vitro* with the plant lectin, concanavalin A, insulin receptors appeared on the lymphocytic surface. In this study a mixed cell population was used which changed continuously during the incubation. In addition, lectins change resting lymphocytes into a new lymphoblastic cell type and do not merely stimulate cell division. Due to this change in cell population, the role of mitosis and/or cell growth in the modulation of receptors for insulin is unclear. In summary, Krug *et al.* (1972) have used a totally different experimental approach from Thomopoulos *et al.* (1977), which makes a comparison between the two studies impossible.

Another reprogramming of the cell which results in a major change in the concentration of insulin receptors is found in lymphocytes exposed to mitogenic stimulation. As we mentioned already, an early intriguing report by Krug *et al.* (1972) indicated that lymphocytes develop an insulin receptor after non-specific mitogen stimulation by the plant lectin, concanavalin A. Later, a series of reports by Helderman and Strom (1977, 1978) clearly demonstrated that unstimulated, freshly prepared thymus-derived (T) and bursal-equivalent (B) lymphocytes do not have insulin receptors. However, the insulin receptor becomes measurable on the lymphocytes after activation by antigen *in vivo* and *in vitro* (Helderman and Strom, 1977, 1978; Helderman *et al.*, 1978), or non-specifically by mitogen (Helderman and Strom, 1978; Helderman *et al.*, 1978). The appearance of the insulin receptor could be explained by at least two mechanisms. First, in the unstimulated lymphocyte, the insulin receptor might be cryptic, in that the receptor structure exists but the hormone receptor interaction cannot occur. A second possible mechanism would be *de novo* synthesis of the insulin receptor, or of a molecule necessary for its binding activity. Based on the observation that phospholipase C, which was previously shown to reveal cryptic insulin receptors in certain cells (Cuatrecasas, 1971), fails to uncover insulin binding sites on unstimulated lymphocytes, Krug *et al.* (1972) suggested that the appearance of insulin receptors after mitogenic stimulation reflects *de novo* synthesis of the insulin receptor, or of other molecules important for the binding capacity of the receptor. The same report suggested that the emergence of the insulin receptor coincided with DNA synthesis, and implied that a causal relationship between the two phenomena existed. However, reports by Helderman and Strom indicate that the acquisition of the hormone receptor is a distinct event from DNA synthesis after cellular activation, indeed, the T-cell insulin receptor becomes measurable after 24 hours of alloantigenic stimulation in the mixed lymphocyte culture, which is much earlier than discernible DNA synthesis in both mouse (Strom *et al.*, 1978) and man (Helderman and Strom, 1978b). The authors therefore conclude that T-cell receptor appearance is a premitotic event. Further, in a more recent study, Helderman and Strom (1979) directly investigated the role of DNA, RNA and protein synthesis in the appearance of the T-lymphocyte insulin receptor after cellular activation. In brief, they showed that blastic transformation is not required for receptor appearance, and that DNA-dependent RNA synthesis followed by protein synthesis results in the development of the insulin binding capacity of the lymphocyte. Based on these observations, they conclude that cellular activation causes *de novo* synthesis of the insulin receptor (or substances crucial to the integrity of the receptor) on the T-lymphocyte rather than uncovers preformed structures (Helderman and Strom, 1979).

Another change in concentration of insulin receptors related to cell differentiation is found in the process of maturation of reticulocytes. Thus, during the maturation of reticulocytes to erythrocytes, the number of insulin binding sites per cell falls from 1600–2000 to 110–260 (Thomopoulos *et al.*, 1978). This difference cannot be accounted for by the twofold increase in erythrocyte volume that occurs in the severe phenylhydrazine anemia used in the animal study as the experimental model for a condition rich in reticulocytes. Similar to the decrease in insulin receptors during the matura- tion of reticulocytes, Friend erythroleukemia cells* show an important reduction in the number of their insulin receptors during the process of differentiation induced by dimethylsulfoxide (Ginsberg *et al.*, 1979). Consist- ent with this are earlier studies by Germinario *et al.* (1977) showing a decrease in glucose transport during the erythroid differentiation. Flow microcytofluorometric measurements have demonstrated an important increase in fluorescent anisotropy during differentiation, suggesting that the membrane becomes more rigid (Arndt-Jovin *et al.*, 1976). This observation, taken together with their preliminary data showing that an increment in membrane fluidity is associated with an increase in the number of insulin receptors, led Ginsberg *et al.* (1979) to suggest that the decrease in membrane fluidity accompanying differentiation may represent the signal for a decrease in insulin receptor concentration.

One of the best-documented examples of cellular reprogramming associated with a change in the concentration of insulin receptors is found in the process of differentiation of the 3T3-L1 cells. Several cloned lines of Swiss mouse embryo 3T3 fibroblasts capable of differentiating into adipocyte-like cells *in vitro*, have been established by Green *et al.* (Todaro and Green, 1963; Green and Kehinde, 1974, 1975, 1976; Green and Meuth, 1974). During the exponential growth the cells seem indistinguishable from previously established 3T3 lines. However, when the cells reach confluence, some cells acquire spontaneously an adipocyte-like morphology. This development of the fatty phenotype is concomitant with a marked increase in *de novo* lipogenesis (Green and Kehinde, 1976), increased activity of lipogenic enzyme (Mackall *et al.*, 1976; Kuri-Harcuch and Green, 1977), and lipopro- tein lipase (Wise and Green, 1978; Spooner *et al.*, 1979). The cells also develop the ability to synthesize cyclic AMP in response to ACTH and show

---

* The Friend erythroleukemia cells (FLC), obtained from a murine leukemia, differentiate upon induction *in vitro* from a cell similar to a primitive erythrocyte
  precursor, the basophilic erythroblast, to a cell resembling a more differentiated
  cell, the orthochromic normoblast (Friend *et al.*, 1971). This differentiation *in vitro*
  seems to reflect the natural process of erythropoiesis. Thus, the process of differentiation of FLC indicates the induction of the enzymes involved in heme synthesis,
  the synthesis of the alpha and beta globin, and the induction of transferrin receptors
  and erythrocyte membrane antigens (Sassa, 1976; Ross *et al.*, 1974; Scher *et al.*,
  1971).

an increased sensitivity to β-adrenergic-mediated stimulation of adenylate cyclase (Rubin *et al*., 1977). Further, with differentiation, there is an increase in insulin binding and the development of sensitivity to the metabolic effects of physiological insulin concentrations (Rubin *et al*., 1977; Reed *et al*., 1977; Karlsson *et al*., 1979). The adipose conversion can happen spontaneously within two to four weeks after the cells have reached confluence, but the degree and rate of the conversion can be enhanced significantly by exposure of the confluent fibroblasts to a variety of agents, such as high concentration of serum (20–30%) (Green and Meuth, 1974), insulin (Green and Kehinde, 1975), prostaglandin $F_{2\alpha}$ (Russell and Ho, 1976), 1-methyl-3-isobutyl-xanthine (MIX) (Russell and Ho, 1976), biotin (Mackall *et al*., 1976), glucocorticoids (Rubin *et al*., 1978), indomethacin (Williams and Polakis, 1977) and finally, the combination of dexamethasone with MIX (Rubin *et al*., 1978). In contrast, epinephrine, dibutyryl-cAMP, and prostaglandin E_1 inhibit the fatty conversion, while bromodeoxyuridine completely blocks the process (Green and Meuth, 1974).

Different laboratories, including our own, have shown that the growth and differentiation of the 3T3-L1 cells are accompanied by marked changes both in the insulin receptors and in the insulin sensitivity of the cells. Thus, the lowest insulin binding is found in the sparse, growing fibroblasts. Upon reaching confluence, there is a fourfold increase in insulin binding, and with differentiation into adipocyte-like cells, a further increase in insulin binding is found. Scatchard analysis of the binding data demonstrates that these increments in insulin binding are accounted for by an augmentation of the number of insulin receptors per cell without significant change in receptor affinity (Karlsson *et al*., 1979). The estimates of the number of insulin receptors reported by various laboratories in general agree well with one another, given the differences in techniques used to measure insulin binding and to analyse the binding data. In the study by Reed *et al*. (1977) the undifferentiated, confluent fibroblasts have 30 000 receptors per cell, while 175 000 receptors per cell are found in fatty fibroblasts. Similarly, Karlsson *et al*. (1979) reported the following number of insulin receptors per cell: 9000 in sparse, growing fibroblasts; 40 000 in confluent, undifferentiated fibroblasts, and 120 000 in fatty fibroblasts. A comparison of the data published by Karlsson *et al*. (1979) and Reed *et al*. (1977) with the numbers reported by Rubin *et al*. (1977) and by Chang and Polakis (1978) is complicated by the fact that the latter two groups have assumed the existence of two independent insulin binding sites, a high affinity–low capacity, and a low affinity–high capacity component. Thus, when the fibroblast differentiates to a fatty fibroblast the number of high affinity insulin receptors increases from 7000 to 250 000, according to the report by Rubin *et al*. (1977), and from 20 000 to 56 000, according to the study by Chang and Polakis (1978).

Reed *et al*. (1977) have demonstrated that the increase in insulin receptors occurring with differentiation cannot be explained by an augmented cell size and surface area due to accumulation of triglycerides. Compared with the non-differentiated 3T3-L1 cells, the differentiated cells have an average cell volume and a surface area which is four times and 2.5 times larger, respectively.

In contrast to the insulin receptor, the number of epidermal growth factor receptors per cell does not change during the differentiation. Therefore, the increase in insulin receptors observed with the adipose conversion is specific for this receptor and does not simply reflect an increased surface area of exposed plasma membranes. Further, using the density-shift method*, Reed and Lane (1980) have recently shown that the rise in insulin binding capacity results from new receptor synthesis.

At present, the regulation of the adipose conversion is not well understood, although the importance of some factors has recently been clarified. Thus, biotin plays an important role in adipose conversion owing to its effects on enzymatic carboxylation (Kuri-Harcuch *et al*., 1978). Without biotin the increased fatty acid synthesis usually associated with the conversion cannot happen, although enzymatic differentiation is not prevented. Therefore, biotin cannot be considered as a regulating factor in the overall process of differentiation. However, one factor has been found to be essential for all aspects of the conversion, and seems to function as a regulator of the overall process of differentiation. The so-called 'adipogenic' factor is a non-lipid factor occurring in most sera studied with the exception of cat serum (Kuri-Harcuch and Green, 1978). The term 'adipogenic' factor has been chosen by Kuri-Harcuch and Green (1978) to distinguish the factor from other substances with a more direct effect on the synthesis of lipids. For example, in the presence of insulin, a faster accumulation of triglycerides and an increased activity of some lipogenic enzymes occurs, demonstrating the lipogenic action of insulin. However, the absence of insulin does not prevent the adipose conversion. In contrast, insulin fails to produce the differentiation when the adipogenic factor is virtually absent (i.e. when medium enriched with cat serum is used). This suggests that the adipogenic factor acts probably at an earlier step than insulin, and appears to be essential for the complete differentiation program.

An unexpected finding reported by our laboratory (Karlsson *et al*., 1979)

---

* In this method cells are cultured in the presence of 'heavy' (^{2}H, ^{13}C, and ^{15}N) amino acids. Isopycnic banding on caesium chloride gradients allows the separation of the solubilized, newly-synthesized 'heavy' and old 'light' receptors. The binding sites for insulin on the 'light' and 'heavy' receptors are then quantitated by a binding assay using ^{125}I-insulin and polyethyleneglycol to precipitate the receptor-bound ^{125}I-insulin. Note that this method was developed to analyse the metabolism of the acetylcholine receptors (Fambrough and Devreotes, 1976; Devreotes *et al*., 1977).

and by Rubin *et al*. (1978) is the absence of insulin-mediated receptor loss ('down-regulation') in 3T3-L1 fatty fibroblasts. At the present time, it is not known whether this lack of insulin's effect on its receptors is due to the cells' failure to respond to the hormonal effect, or to regulation of receptor synthesis and degradation induced by insulin itself, resulting in a constant receptor concentration.

Recently, Négrel *et al*. (1978) and Grimaldi *et al*. (1979) described another cell line (ob-17), whose cells differentiate to cells having the morphological and biochemical characteristics of adipocytes. This clonal cell line, designated ob-17, was derived from the epididymal fat pad of the genetic ob/ob mouse (C57BL/6J). Similarly to the 3T3-L1 cells, after growth arrest, the ob-17 cells differentiate into adipocyte-like cells, and increase the number of their insulin receptors. In contrast to the 3T3-L1 fatty fibroblasts, the ob-17 adipose cells can down-regulate the number of the insulin receptors after prolonged exposure to insulin. This loss of the 'high affinity' binding sites is accompanied by the disappearance of the stimulatory effect of insulin on α-aminoiso-butyrate uptake. It is worth mentioning that this insulin resistance for α-aminoisobutyrate transport by the differentiated ob-17 cells is quite analogous to the insulin resistance for glucose transport encountered in mature adipocytes during hyperinsulinemia *in vivo* (Kobayashi and Olefsky, 1978).

2.6.3 Down-regulation

The hormone itself plays a crucial role in the regulation of the concentration of its receptor. Indeed, an inverse relationship between the serum concentration of insulin and the concentration of cell membrane-insulin receptors has been clearly demonstrated in a variety of animal and human disease models (Kahn, 1976; Roth, 1979a). The most convincing evidence that insulin exerts a direct effect on its own receptor comes from experiments performed by Gavin *et al*. (1974) and Kosmakos and Roth (1980). When cultured human lymphocytes (IM-9 line), which had reached a stationary phase of growth, were exposed to varying concentrations of insulin, extensively washed, and then studied for ^{125}I-insulin binding, a decrease in the concentration of insulin receptors was observed. Both the rate of onset of the loss and the steady-state new concentration of receptors are a direct function of the concentration of ambient insulin (Fig. 2.5). The lowered steady-state level is maintained while insulin is present. There appears to be a plateau beyond which receptor number does not decrease. Thus, insulin cannot lower the number of receptors by more than 60–85%. Note that the sensitivity of the insulin receptor to the hormone regulatory effect varies from one system to the other (Fig. 2.6). Analogs of insulin, which vary over a 100-fold range in biological potency and affinity for

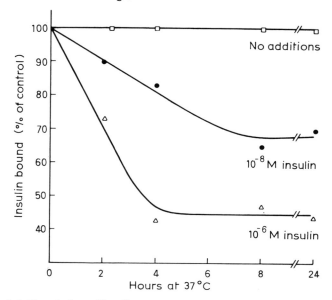

Fig. 2.5 Regulation of insulin receptors by insulin. Human lymphoblastoid cells (IM-9) which exist in continuous culture were cultured for up to 16 hours in the presence of the indicated insulin concentrations. At the indicated times, cells were extensively washed, and the insulin receptor concentration was determined. Receptor concentration showed a time- and dose-dependent fall in cells which had been exposed to insulin. Modified from Gavin *et al.* (1974).

receptor, exert their regulatory effects on the insulin receptor in direct proportion to their bioactivities. Further, the insulin analog desalanine-desasparagine insulin, which is unable to induce negatively cooperative site–site interactions among insulin receptors, evokes the regulatory effect. These data suggest that the mechanism of regulation of the insulin receptor concentration and cooperativity are independent, and that the loss of receptor is likely to be mediated through the 'bioactivity' site of the insulin molecule. For insulin-mediated loss of its receptor to occur, insulin has to bind to its receptor, but this condition is not sufficient. Indeed, lowering the temperature from 37° C to 24° C prevents insulin from exerting its regulatory effect, although at the lower temperature the steady-state level of binding of the hormone to its receptor is actually greater than at the higher temperature. Further, inhibition of protein synthesis by cycloheximide and inhibition of cellular energy production by azide and nitrophenol also block the insulin-

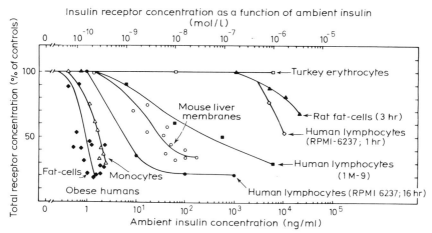

Fig. 2.6 Summary of relationship of receptor concentration to ambient insulin. The relationship of receptor concentration per cell as a function of the ambient insulin concentration is shown here for multiple systems. For obese humans, isolated fat-cells and circulating monocytes were both studied. Insulin receptors on plasma membrane of liver from rats with hyperinsulinemia and hypoinsulinemia fit approximately the same relationship as in humans. With receptors in mice, there is a similar relationship, although over a different range of insulin concentrations (mice *in vivo* have higher insulin concentrations than humans). Human lymphocytes in cell culture have insulin receptors on their cell surface; the receptor concentration at steady state (16–18 hours) is predicted by the concentration of insulin in the medium. Cells that are incubated for shorter time periods show the same relationship but at higher insulin concentrations (rat fat-cells incubated for 3 hours or human lymphocytes incubated for 1 hour). Turkey erythrocytes, incubated for up to 24 hours at 30° C in the presence of insulin, showed no change in the concentration of their insulin receptor. From Roth, J. (1979a).

induced loss of receptors (Kosmakos and Roth, 1980). After removal of insulin from the medium, the concentration of insulin receptors increases rapidly. Thus, within six hours, approximately half of the lost receptors are regenerated, and normal levels are found within 18 to 24 hours. The recovery process is very sensitive to insulin. Indeed, as little as 10^{-9} M-insulin partially prevents the restoration of the full complement of receptors (Gavin *et al.*, 1974). The return of the insulin receptors is blocked by cycloheximide, suggesting that synthesis of receptor molecules is necessary for the recovery of insulin receptors.

A remarkable characteristic of the hormone-mediated receptor loss is its

exquisite specificity. Thus, in cultured human lymphocytes, insulin induces
a decrease of its cell surface receptors, but fails to affect the growth
hormone receptors on the same cell (Gavin *et al.*, 1974). Conversely,
human growth hormone induces a loss of its receptors without interfering
with the insulin receptors (Lesniak and Roth, 1976).

The demonstration *in vitro* of the insulin-mediated receptor loss is not
limited to cultured human lymphocytes. Thus, insulin at concentrations comparable to those found *in vivo* (10^{-8}–10^{-9} M) induces an approximately 50
per cent loss of insulin receptors in primary monolayer cultures of adult rat
hepatocytes (Blackard *et al.*, 1978) and human diploid fibroblasts (Mott *et al.*,
1979). Similarly, chronic exposure to insulin results in a significant decrease
in insulin receptor in ob-17 adipocytes, cells of a clonal cell line (ob-17)
isolated from the epididymal fat of ob/ob mouse (Grimaldi *et al.*, 1979), and
in 3T3-L1 fibroblasts, cells of a clonal cell-line (3T3-C2) isolated from dis-
aggregated mouse embryo (Chang and Polakis, 1978; Rubin *et al.*, 1977;
Reed *et al.*, 1977).

The self-regulation of membrane receptors is not unique for the insulin
receptor. Originally described with surface antigens, the phenomenon has
been observed in cells exposed to a wide variety of ligands such as glycoprotein hormones, neurotransmitters, and surface-modulating factors (e.g. lectin,
immunoglobulin) (Raff, 1976; Catt *et al.*, 1979). A negative feedback regulation by homologous hormones and growth factors on their own receptors has
been demonstrated for hGH (Lesniak and Roth, 1976; Lesniak *et al.*, 1977),
thyrotropin-releasing hormone (Hinkle and Tashijian, 1976), gonadotropins
(Catt and Dufau, 1977), catecholamines (Mukherjee *et al.*, 1975) and
glucagon (Srikart *et al.*, 1977). In contrast, exposure of cells to elevated
concentrations of prolactin (Posner *et al.*, 1975) and angiotensin II (Hauger *et
al.*, 1978) will increase the number of their homologous receptors.

The mechanism of the insulin-mediated loss is not known at the present
time. Using a radioimmunoassay of the insulin receptor, it was shown that the
loss of insulin binding sites is associated with a loss in immunoreactive
receptors (Harrison *et al.*, 1979), indicating that down-regulation is not due to
inactivation of the insulin binding function *in situ*, but rather to a disappearance of the receptor from the cell membrane. This disappearance of
the insulin receptor from the cell surface may be due to shedding of the
receptor molecule into the medium, or to destruction of the receptor at the
level of the cell membrane itself, or within the cell following internalization.
Different observations provide indirect evidence that the latter mechanism is
the most plausible. First, quantitative radioautographic electron microscopic
studies indicate that there is a relationship between the concentration of
hormone or growth factor needed to induce receptor loss and the extent and
rate of internalization of the ligand (Gorden *et al.*, 1980); second, adsorptive
pinocytosis increases during insulin-induced receptor loss (Gorden *et al.*,

1979) and, finally, insulin has been shown to form clusters, patches and even caps (Schlessinger *et al.*, 1978, 1980; Schlessinger, 1980; Barazzone *et al.*, 1980). These cell-membrane phenomena, reflecting existing or induced microheterogeneity of the membrane, would provide the cell with an appropriate mechanism for removing specifically a single type of its own receptors from the cell surface.

2.6.4 Negative cooperativity

Insulin itself can regulate the affinity of its own receptor. The Scatchard plot (plot of bound/free hormone *versus* bound hormone) for insulin binding is curvilinear with upward concavity. This was previously thought to reflect the presence of two or more distinct classes of independent sites with different affinities for insulin (Kahn *et al.*, 1974). However, in 1973, De Meyts *et al.* suggested that a more likely explanation for this non-Michaelian binding isotherm was the existence of 'negative cooperativity' among the receptor sites, i.e. the average affinity of the insulin receptors progressively decreases as the occupancy of the sites by insulin increases. The presence of negative cooperativity was kinetically demonstrated by the following experiment (Fig. 2.7). Cultured human lymphocytes (IM-9 line) were incubated in a single batch with ^{125}I-insulin. The experimental conditions were such that only a small portion of the receptor sites was occupied by the tracer. After being spun down to allow for removal of the unbound tracer, the cells were resuspended in the original volume. Aliquots of the cells were then immediately diluted (100-fold) into two series of tubes; one series contained

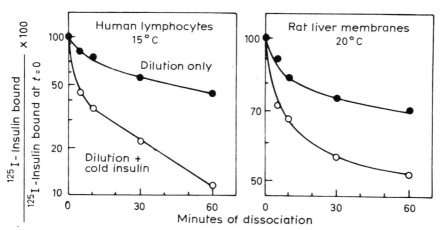

Fig. 2.7 Kinetic experiment to test for negative cooperativity in human cultured lymphocytes and rat liver membranes. From De Meyts *et al.* (1976).

only buffer ('dilution alone'), while the second was supplemented with an excess of unlabeled ('dilution and unlabeled insulin'). Thereafter, the dissociation of the prebound ^{125}I-insulin was followed as a function of time. Note that the dilution was so great, theoretically and experimentally, that rebinding of the dissociated tracer was prevented. The authors reasoned that if the insulin binding sites were independent, the dissociation rate of the labeled insulin should not be altered by the presence of the unlabeled hormone, which, at the high concentration, quickly occupies the empty sites. In fact, as predicted by the negative cooperativity model, the dissociation rate of ^{125}I-insulin was markedly enhanced in the presence of unlabeled hormone. Care was taken by De Meyts *et al.* (1973, 1976) to exclude the counterhypothesis that the accelerated dissociation of ^{125}I-insulin observed in the presence of unlabeled insulin was due to less rebinding, e.g. due to inadequate dilution, unstirred layer, insufficient mixing. In contrast, when an identical experiment was performed with human growth hormone, the unlabeled hormone did not affect the dissociation rate of the prebound tracer, suggesting the existence of one class of independent binding sites. This is totally consistent with the linear Scatchard plot observed with hGH. The kinetic demonstration of negative cooperativity developed by De Meyts *et al.* was later successfully extended by various groups to other ligands binding to cell surface receptors and yielding curvilinear Scatchard plots (e.g. nerve growth factor, epidermal growth factor, thyroid-stimulating hormone, thrombin, ACTH and lectins) (De Meyts, 1976). For insulin, the phenomenon of negative cooperativity can be demonstrated not only on whole cells and membranes, but also on detergent-solubilized receptors, suggesting that it is an intrinsic property of the receptor rather than the result of membrane-mediated alterations.

The negative cooperativity of the insulin receptor appears to be a very fundamental property of the receptor and of insulin. Indeed, it was found to occur in the receptors of all species examined and to be generated by all natural insulins tested (De Meyts *et al.*, 1978). Further, extensive studies with natural insulins and insulin-like molecules as well as chemically altered insulins which differ over a 1000-fold range in affinity for receptor and in biopotency, suggested that the residues involved in inducing the negative cooperativity compose a distinct subsite (i.e. a cooperative site) within the binding region. Further strong support for the concept of negative cooperativity has recently been provided by De Meyts (1980) with studies of the binding properties of the non-cooperative insulin analog, desalanine-desasparagine (DAA) insulin. Thus, unlabeled DAA-insulin fails to provoke the accelerated dissociation of either labeled insulin or labeled DAA-insulin, although DAA does bind to the insulin receptor and does stimulate the characteristic biological effects of insulin. In contrast, unlabeled insulin increases the dissociation rate of prebound ^{125}I-labeled DAA-insulin. These data support the idea that

DAA-insulin binds to the receptor in a high-affinity state, but is unable to evoke site–site interactions that result in decreased receptor affinity. However, the affinity of the receptors occupied by DAA-insulin can be reduced by site–site interaction generated by insulin bound to neighbouring sites. Furthermore, the Scatchard plot of DAA-insulin binding is linear and reveals that DAA-insulin binds to the same number of sites as does native insulin.

De Meyts and Roth (1975) have proposed a specific graphic analysis for binding data of the insulin receptor in which there are at least two interconvertible states: (i) a 'high affinity' state in which insulin dissociates slowly ('empty sites' conformation) and with the binding constant $\bar{K}_e$, and (ii) a 'low affinity' state in which insulin dissociates fast ('filled sites' conformation) and with a binding constant $\bar{K}_f$. In the absence of insulin, the receptors are entirely in the high affinity form. As receptor sites become occupied, a portion of the receptors converts to the low affinity form. The proportion of sites in each state can be affected by a number of factors summarized in Table 2.6.

The negative cooperativity model, which proposes that receptor affinity is inversely related to receptor occupancy, predicts that when the receptors are pre-loaded with a high concentration of insulin, then dissociation from the receptor will be accelerated upon dilution only, even in the absence of cold insulin in the medium. When De Meyts carried out such experiments (De Meyts *et al.*, 1976), he found that pre-loading of the receptor, followed by

Table 2.6 'Two-state' model for insulin receptors

'Slow-dissociating state' favored by:	'Fast-dissociating' state favored by:
–Low fractional saturation with insulin	–Increased fractional saturation with insulin
–DAA insulin	–Acid pH (5–6)
–Insulin at high concentration ($\geqslant 10^{-7}$ M) self-associates and concomitantly its ability to induce negative cooperativity decreases as dimerization increases	–High temperature (37°C)
	–Low concentration of Ca^{2+}/Mg^{2+}
	–Urea*
–Alkaline pH (8–9)	–Anti-receptor antibodies
–Low temperature (4°C)	
–High concentration of Ca^{2+}/Mg^{2+}	
–Concanavalin A (20 μg/ml)	

* The fact that urea can promote almost complete dissociation of insulin from receptors within minutes indicates that no covalent bonds are involved in insulin binding or in the mechanisms regulating insulin dissociation. Direct effects of urea on the insulin–receptor interaction cannot, however, be distinguished from possible effects on intersubunit bonds in the receptor.

dilution only, gave results exceedingly close to those obtained by loading the receptor with a tracer only, followed by dilution in the presence of high concentrations of unlabeled insulin. When Pollet repeated these experiments, he obtained different results, namely, dissociation of tracer insulin on dilution only was independent of the concentration of insulin that had been pre-bound to the receptor (Pollet *et al.*, 1977). It should be noted that Pollet did reproduce the curvilinear Scatchard plots for steady-state insulin binding and also reproduced the characteristic difference in dissociation rates when he used only a tracer of ^{125}I-insulin prior to commencement of the dissociation. Pollet interpreted his experiments to indicate that steady-state insulin binding gave curvilinear Scatchard plots because there were two discrete classes of sites, each with a fixed affinity, and that the dissociation kinetics were explained by a concerted exchange mechanism in which the unlabeled insulin accelerated the dissociation of the bound labeled insulin. De Meyts has pointed out that he was able to reproduce fully his own results and, in addition, could reproduce Pollet's results simply by delaying the onset of measurements of dissociation for a few minutes after the commencement of the dissociation. Rodbard and De Lean, in their analysis of Pollet's data, indicate that, in fact, the delay of a few minutes in the onset of such measurements could reproduce these findings (De Meyts, 1980).

In summary, the crux of the dispute between De Meyts and Pollet is focused entirely on one experiment in which the two investigators obtained opposite results, one totally consistent with the negative cooperativity model, and the other inconsistent with it. More importantly, it should be emphasized that Pollet's interpretation requires two separate explanations for the steady-state and kinetic findings, whereas the negative cooperativity model explains both the steady-state and the kinetic data. Further, it should be emphasized that the negative cooperativity model provides a consistent single interpretation for a very wide range of findings, including linearization of the Scatchard plot in the high-affinity state or linearization of the Scatchard plot in the low-affinity state (by a wide range of diverse mechanisms and conditions). In each of these conditions, the kinetics of dissociation by dilution only or by dilution in the presence of excess of unlabeled insulin provides data that are consistent with the steady-state observations. None of these diverse perturbations, which are satisfactorily explained by the negative cooperativity model, would fit with the two separate mechanisms provided by Pollet's single experiment. Thus, for the present, we conclude that there is an extraordinarily large body of experimental data under a wide range of steady-state and kinetic conditions that supports the negative cooperativity model.

2.6.5 Temperature, Ions

An important factor affecting insulin binding to its cellular receptors is temperature. Thus, at higher temperature, the initial rate of insulin binding is increased and the steady-state level of binding is reached faster. However, the actual steady-state level of binding is inversely related to the temperature (Ginsberg, 1977). This phenomenon has been observed for almost all insulin receptors and is mediated by a greater increment in the dissociation rate than in the association rate with increasing temperature. Note that both degradation of hormone and degradation of receptors are extremely important factors that interfere with precise measurements of the temperature dependence of steady-state binding. Different recent biochemical and morphological studies have clearly demonstrated that at physiological temperatures (37° C), insulin is rapidly internalized following initial binding to its specific cell surface receptor. Further, the internalized labeled material associates preferentially with lysosome-like structures. It is thus evident that the binding of insulin to cells at physiological temperature is not a simple reversible interaction, and this caution should be taken into consideration in the analysis of binding data.

As anticipated for a protein–ligand interaction, the affinity of insulin for its receptor is influenced by the ionic environment. For most cells the pH optimum for insulin binding lies between 7.8 and 8.0 (Roth, 1979a; Ginsberg, 1977). The insulin receptor is extremely sensitive to pH. Indeed, below 7.0 and above 8.2, binding usually decreases to less than half. It should be stressed that the pH optimum of insulin binding is markedly dependent on the temperature, as shown by Waelbroeck *et al.* (1979). One major pathophysiological implication of the pH dependence of insulin binding relates to the insulin resistance in diabetic ketoacidosis, where a drop in the pH of extracellular fluid from 7.4 to 6.8 might be responsible for an important decrease in insulin binding consequent to the reduced affinity.

Other aspects of the ionic environment appear to have a variable role in the binding of insulin, depending on the receptor system. Thus, Ca^{2+} and Mg^{2+} significantly increase insulin binding to its receptors on human placental membranes and avian erythrocytes (Posner, 1974; Ginsberg, 1977), whereas these ions fail to affect insulin binding to cultured human lymphocytes (Gavin *et al.*, 1973) and inhibit insulin binding to cultured human placental cells (Podskalny *et al.*, 1975). Further, high ionic strength (2 M − Na^+) markedly enhances insulin binding to liver and fat-cell membranes. A decrease in insulin degradation and an increase in insulin receptor affinity are responsible for this phenomenon (Kahn *et al.*, 1974).

REFERENCES

Arndt-Jovin, D.J., Ostertag, W., Wisen, H., Klimek, F. and Jovin, T. (1976), *J. Histochem. Cytochem.*, **24**, 332–347.

Bar, R.S., Koren, H. and Roth, J. (1976), *Diabetes*, **25**, Suppl. I, 348.

Barazzone, P., Carpentier, J.-L., Gorden, P., Van Obberghen, E. and Orci, L. (1980), *Nature*, **286**, 401–403.

Bergeron, J.J.M., Evans, W.H. and Geschwind, I.I. (1973), *J. Cell Biol.*, **59**, 771–776.
Bergeron, J.J.M., Sikstrom, R., Hand, A.R. and Posner, B.I. (1979), *J. Cell Biol.*, **80**, 427–443.
Blackard, W.G., Guzelian, P.S. and Small, M.E. (1978), *Endocrinology,* **103**, 548–554.
Carpentier, J.-L., Gorden, P., Amherdt, M., Van Obberghen, E., Kahn, C.R. and Orci, L. (1978), *J. Clin. Invest.*, **61**, 1057–1070.
Carpentier, J.-L., Gorden, P., Freychet, P., LeCam, A. and Orci, L. (1979), *J. Clin. Invest.*, **63**, 1249–1261.
Carpentier, J.-L., Van Obberghen, E., Gorden, P. and Orci, L. (1980), *Diabetologia*, **19**, 263.
Catt, K.J. and Dufau, M.L. (1977), *Annu. Rev. Physiol.*, **39**, 529–557.
Catt, K.J., Harwood, J.P., Aguilera, G. and Dufau, M.L. (1979), *Nature*, **280**, 109–116.
Chang, T.H. and Polakis, S.E. (1978), *J. Biol. Chem.*, **253**, 4693–4696.
Cuatrecasas, P. (1971), *J. Biol. Chem.*, **246**, 6532–6542.
Cuatrecasas, P. (1972), *J. Biol. Chem.*, **247**, 1980–1991.
Cuatrecasas, P. (1974), *Annu. Rev. Biochem.*, **43**, 169–214.
Cuatrecasas, P. and Tell, G.P.E. (1973), *Proc. Natl. Acad. Sci. USA*, **71**, 485–489.
De Meyts, P. (1976), in *Methods in Receptor Research* (Blecher, M., ed), Marcel Dekker, New York, pp. 301–383.
De Meyts, P. (1976), *J. Supramolec. Struct.*, **4**, 241–258.
De Meyts, P. (1980), in *Hormones and Cell Regulation* (Dumont, J. and Nunez, J., eds), in press.
De Meyts, P. and Roth, J. (1975), *Biochem. Biophys. Res. Commun.*, **66**, 1118–1126.
De Meyts, P., Roth, J., Neville, D.M., Jr., Gavin, J.R., III and Lesniak, M.A. (1973), *Biochem. Biophys. Res. Commun.*, **55**, 154–161.
De Meyts, P., Bianco, A.R. and Roth, J. (1976), *J. Biol. Chem.*, **251**, 1877–1888.
De Meyts, P., Van Obberghen, E., Roth, J., Wollmer, A. and Brandenberg, D. (1978), *Nature,* **273**, 504–509.
Desbuquois, B. and Aurbach, G.D. (1971), *J. Clin. Endocrinol. Metab.*, **33**, 732–738.
Devreotes, P.N., Gardner, J.M. and Fambrough, D.M. (1977), *Cell*, **10**, 365–373.
Fambrough, D.M. and Devreotes, P.N. (1976), in *Biogenesis and Turnover of Membrane Macromolecules* (Cook, J.S., ed.), Raven Press, New York, pp. 124–144.
Flier, J.S., Kahn, C.R., Roth, J. and Bar, R.S. (1975), *Science*, **190**, 63–65.
Freychet, P., Roth, J. and Neville, D.M., Jr., (1971), *Proc. Natl. Acad. Sci. USA*, **68**, 1833–1837.
Freychet, P., Brandenburg, D. and Wollmer, A. (1975), *Diabetologia*, **10**, 1–5.
Friend, C., Scher, W., Holland, J.G. and Sato, T. (1971), *Proc. Natl. Acad. Sci. USA*, **68**, 378–382.
Gavin, J.R., III, Mann, D.L., Buell, D.N. and Roth, J. (1972), *Biochem. Biophys. Res. Commun.*, **49**, 870–876.
Gavin, J.R., III, Gorden, P., Roth, J., Archer, J.A. and Buell, D.N. (1973), *J. Biol. Chem.*, **248**, 2202–2207.
Gavin, J.R., III, Roth, J., Neville, D.M., Jr., De Meyts, P. and Buell, D.N. (1974), *Proc. Natl. Acad. Sci. USA*, **71**, 84–88.

Germinario, R.H., Kleiman, L., Peters, S. and Olivera, M. (1977), *Exp. Cell Res.*, 110, 375–385.

Ginsberg, B.H. (1977), in *Biochemical Actions of Hormones* (Litwak, G., ed.), Vol. 4, Academic Press, New York, pp. 313–349.

Ginsberg, B.H., Kahn, C.R., Roth, J. and De Meyts, P. (1976), *Biochem. Biophys. Res. Commun.*, 73, 1068–1074.

Ginsberg, B.H., Brown, T. and Raizada, M. (1979), *Diabetes*, 28, 823–827.

Gliemann, J. and Gammeltoft, S. (1974), *Diabetologia*, 10, 105–113.

Goldfine, I.D. and Smith, G.J. (1976), *Proc. Natl. Acad. Sci. USA*, 73, 1427–1431.

Goldfine, I.D., Jones, A.L., Hradek, G.T., Wong, K.Y. and Mooney, J.S. (1978), *Science*, 202, 760–763.

Goldstein, J.L., Anderson, R.G.W. and Brown, M.S. (1979), *Nature*, 279, 679–684.

Goodfriend, T.L. and Lin, S.-Y. (1970), *Circ. Res.*, 26/27, I-163–I-174, Suppl. 1.

Gorden, P., Carpentier, J.-L., Freychet, P., LeCam, A. and Orci, L. (1978), *Science*, 200, 782–785.

Gorden, P., Carpentier, J.-L., Van Obberghen, E., Barazzone, P., Roth, J. and Orci, L. (1979), *J. Cell Sci.*, 39, 77–88.

Gorden, P., Carpentier, J.-L., Freychet, P. and Orci, L. (1980), *Diabetologia*, 18, 263–274.

Green, H. and Kehinde, O. (1974), *Cell*, 1, 113–116.

Green, H. and Kehinde, O. (1975), *Cell*, 5, 19–27.

Green, H. and Kehinde, O. (1976), *Cell*, 7, 105–113.

Green, H. and Meuth, M. (1974), *Cell*, 3, 127–131.

Grimaldi, P., Négrel, R., Vincent, J.P. and Ailhaud, G. (1979), *J. Biol. Chem.*, 254, 6849–6852.

Harmon, J.T., Kahn, C.R., Kempner, E.J. and Schlegel, W. (1980), *J. Biol. Chem.*, 255, 3412–3419.

Harrison, L.C. and Kahn, C.R. (1980), *Prog. Clin. Immunol.*, 4, 107–125.

Harrison, L.C., Flier, J.S., Itin, A., Kahn, C.R. and Roth, J. (1979), *Science*, 203, 544–547.

Hauger, R.L., Aguilera, G. and Catt, K.J. (1978), *Nature*, 271, 176–177.

Helderman, J.H. and Strom, T.B. (1977), *J. Clin. Invest.*, 59, 338–344.

Helderman, J.H. and Strom, T.B. (1978a), *Nature*, 274, 62–63.

Helderman, J.H. and Strom, T.B. (1978b), *Abstr. 7th Int. Congr. Nephr.*, J-1.

Helderman, J.H. and Strom, T.B. (1979), *J. Biol. Chem.*, 254, 7203–7207.

Helderman, J.H., Reynolds, T.C. and Strom, T.B. (1978), *Eur. J. Immunol.*, 8, 589–595.

Hinkle, P.M. and Tashijian, A.H. Jr. (1976), *Biochemistry*, 14, 3845–3851.

Horvat, A., Li, E. and Katsoyannis, P.G. (1975), *Biochim. Biophys. Acta*, 382, 609–620.

Jarett, L. and Smith, R.M. (1977), *J. Supramol. Struct.*, 6, 45–59.

Jarrett, D.B., Roth, J., Kahn, C.R. and Flier, J.S. (1976), *Proc. Natl. Acad. Sci. USA*, 73, 4115–4119.

Kahn, C.R. (1976), *J. Cell Biol.*, 70, 261–286.

Kahn, C.R., Neville, D.M., Jr., and Roth, J. (1973), *J. Biol. Chem.*, 248, 244–250.

Kahn, C.R., Freychet, P., Neville, D.M., Jr. and Roth, J. (1974), *J. Biol. Chem.*, 249, 2249–2257.

Kahn, C.R., Flier, J.S., Bar, R.S., Archer, J.A., Gorden, P., Martin, M.M. and Roth, J. (1976), *N. Engl. J. Med.*, **294**, 739–745.

Karlsson, F. A., Grunfeld, C., Kahn, C.R. and Roth, J. (1978), *Endocrinology*, **104**, 1383–1392.

Karlsson, F.A., Grunfeld, C., Kahn, C.R. and Roth, J. (1979), *Program and Abstracts 60th Annual Meeting of the Endocrine Society*, p. 161, abstract 174.

Kasuga, O., Van Obberghen, E., Yamada, K. and Harrison, L.C. (1981), *Diabetes.*, in press.

Kobayashi, M. and Olefsky, J.H. (1978), *J. Clin. Invest.*, **62**, 73–81.

Kono, T. (1971), *J. Biol. Chem.*, **244**, 5777–5784.

Kono, T. and Barham, F.S. (1971), *J. Biol. Chem.*, **246**, 6210–6216.

Kosmakos, F. and Roth, J. (1980), *J. Biol. Chem.*, **20**, 9860–9869.

Krug, U., Krug, F. and Cuatrecasas, P. (1972), *Proc. Natl. Acad. Sci. USA*, **69**, 2604–2608.

Kuri-Harcuch, W. and Green, H. (1977), *J. Biol. Chem.*, **252**, 2158–2160.

Kuri-Harcuch, W. and Green, H. (1978), *Proc. Natl. Acad. Sci. USA*, **75**, 6107–6109.

Kuri-Harcuch, W., Wise, L.S. and Green, H. (1978), *Cell*, **14**, 53–59.

Lang, U., Kahn, C.R. and Chrambach, A. (1979), *Endocrinology*, **106**, 40–49.

Larner, J., Galasko, G., Cheng, K., De Paoli-Roach, A.A., Huang, L., Daggy, P. and Kellogg, J. (1980), *Science*, **206**, 1408–1410.

Lefkowitz, R.J., Pastan, I. and Roth, J. (1969), in *The Role of Adenyl Cyclase and Cyclic 3',5'-AMP in Biological Systems*, (Rall, T.W., Rodbell, M. and Condliffe, P., eds), NIH Fogarty International Center Proceedings, Bethesda, Maryland.

Lefkowitz, R.J., Roth, J., Pricer, W. and Pastan, I. (1980), *Proc. Natl. Acad. Sci. USA*, **65**, 745–762.

Lesniak, M.A. and Roth, J. (1976), *J. Biol. Chem.*, **251**, 3720–3729.

Lesniak, M.A., Gorden, P. and Roth, J. (1977), *J. Clin. Endocrinol. Metab.* **44**, 838–849.

Lin, S.-Y. and Goodfriend, T.L. (1970), *Amer. J. Physiol.*, **218**, 1319–1328.

Mackall, J.C., Student, A.K., Polakis, S.E. and Lane, M.D. (1976), *J. Biol. Chem.*, **251**, 6462–6464.

Maturo, J.M. and Hollenberg, M.D. (1978), *Proc. Natl. Acad. Sci. USA*, **75**, 3070–3074.

Mott, D.M., Howard, B.V. and Bennett, P.H. (1979), *J. Biol. Chem.*, **254**, 8762–8767.

Mukherjee, C., Caron, M.C. and Lefkowitz, R. (1975), *Proc. Natl. Acad. Sci. USA*, **72**, 1945–1949.

Négrel, R., Grimaldi, P. and Ailhaud, G. (1978), *Proc. Natl. Acad. Sci. USA*, **75**, 6054–6058.

Newerly, K. and Berson, S.A. (1957), *Proc. Soc. Exp. Biol. Med.*, **94**, 751–755.

Orci, L., Rufener, C., Malaisse-Lagae, F., Blondel, B., Amherdt, M., Bataille, D., Freychet, P. and Perrelet, A. (1975), *Isr. J. Med. Sci.*, **11**, 639–655.

Pastan, I., Johnson, G.S. and Anderson, W.B. (1975), *Annu. Rev. Biochem.*, **44**, 491–522.

Pilch, P.F. and Czech, M.P. (1980), *J. Biol. Chem.*, **255**, 1722–1731.

Pollet, R.J., Standaert, M.L. and Haase, B.A. (1977), *J. Biol. Chem.*, **252**, 5828–5834.

Podskalny, J.M., Chou, J.Y. and Rechler, M.M. (1975), *Arch. Biochem. Biophys.*, **170**, 504–513.

Posner, B.I. (1974), *Diabetes,* **23**, 209–217.

Posner, B.I., Kelly, P.A. and Friesen, H.G. (1975), *Science,* **188**, 57–59.

Raff, M. (1976), *Nature,* **250**, 265–266.

Reed, B.C. and Lane, D.M. (1980), *Proc. Natl. Acad. Sci. USA,* **77**, 285–289.

Reed, B.C., Kaufman, S.H., Mackall, J.C., Student, A.K. and Lane, M.D. (1977), *Proc. Natl. Acad. Sci. USA,* **74**, 4876–4880.

Rosen, O.M., Chia, G.H., Fung, C. and Rubin, C.S. (1979), *J. Cell Physiol.,* **99**, 37–42.

Ross, J., Gielen, J., Packman, S., Ikawa, Y. and Leder, P. (1974), *J. Mol. Biol.,* **87**, 697–714.

Roth, J. (1973), *Metabolism,* **22**, 1059–1073.

Roth, J. (1979a), in *Endocrinology,* Vol. 3 (De Groot, L., ed.), Grune and Stratton, New York, pp. 2037–2054.

Roth, J. (1979b), in *Developments in Cell Biology,* Vol. 4 (Delisi, C. and Blumenthal, R., eds), Elsevier North Holland, Inc., Amsterdam, pp. 185–196.

Rubin, C.S., Lai, E. and Rosen, O.M. (1977), *J. Biol. Chem.,* **252**, 3554–3557.

Rubin, C.S., Hirsch, A., Fung, C. and Rosen, O.M. (1978), *J. Biol. Chem.,* **253**, 7570–7578.

Russell, T.R. and Ho, R.-J. (1976), *Proc. Natl. Acad. Sci. USA,* **73**, 4516–4520.

Sassa, S. (1976), *J. Exp. Med.,* **43**, 305–315.

Scher, W., Holland, J.G. and Friend, C. (1971), *Blood,* **37**, 428–437.

Schlessinger, J. (1980), *Trends in Biochemical Sciences,* **5**, 210–214.

Schlessinger, J., Schechter, Y., Willingham, M.C. and Pastan, I. (1978), *Proc. Natl. Acad. Sci. USA,* **75**, 2659–2663.

Schlessinger, J., Van Obberghen, E. and Kahn, C.R. (1980), *Nature,* **286**, 729–731.

Seals, J.R. and Jarett, L. (1980), *Proc. Natl. Acad. Sci. USA,* **77**, 77–81.

Spooner, P.M., Chernick, S.S., Garrison, M.M. and Scow, R. (1979), *J. Biol. Chem.,* **254**, 1305–1311.

Srikanrt, C.B., Freeman, D., McCorkle, K. and Unger, R. (1977), *J. Biol. Chem.,* **252**, 7434–7436.

Stadie, W.C., Haugaard, N. and Vaughan, M.J. (1953), *J. Biol. Chem.,* **200**, 745–751.

Strom, T.B., Helderman, J.H. and Williams, R.M. (1978), *Immunogenetics,* **7**, 51–56.

Suzuki, K. and Kono, T. (1979), *J. Biol. Chem.,* **254**, 9786–9794.

Terris, S. and Steiner, D.F. (1975), *J. Biol. Chem.,* **250**, 8389–8398.

Thompoulos, P., Roth, J., Lovelace, E. and Pastan, I. (1976), *Cell,* **8**, 417–423.

Thomopoulos, P., Kosmakos, F.C., Pastan, I. and Lovelace, E. (1977), *Biochem. Biophys. Res. Commun.,* **75**, 246–252.

Thomopoulos, P., Berthellier, M. and Laudat, M.H. (1978), *Biochem. Biophys. Res. Commun.,* **85**, 1460–1464.

Todaro, G.J. and Green, H. (1963), *J. Cell Biol.,* **17**, 299–313.

Wachslicht-Rodbard, H., Zeleznik, A.J., McGuire, E.A., Berman, M., Rodbard, D. and Roth, J., in preparation.

Waelbroeck, M., Van Obberghen, E. and De Meyts, P. (1979), *J. Biol. Chem.,* **254**, 7736–7740.

Williams, I.H. and Polakis, S.E. (1977), *Biochem. Biophys. Res. Commun.,* **77**, 175–186.

Wise, L.S. and Green, H. (1978), *Cell,* **13**, 233–242.

Wisher, M.H., Baron, M.D., Jones, R.H., Sonksen, P.H., Saunders, D.J., Thamm, P. and Brandenburg, D. (1980), *Biochem. Biophys. Res. Commun.,* **92**, 492–498.
Yip, C.C., Yeung, C.W.T. and Moule, M.L. (1978), *J. Biol. Chem.,* **253**, 1743–1745.
Yip, C.C., Yeung, C.W.T. and Moule, M.L. (1980), *Biochemistry,* **19**, 70–76.
Zeleznik, A.J. and Roth, J. (1978), *J. Clin. Invest.,* **61**, 1363–1374.

3 EGF: Receptor Interactions and the Stimulation of Cell Growth

GRAHAM CARPENTER and STANLEY COHEN

Acknowledgements

The authors wish to acknowledge grants CA24071 from the National Cancer Institute (G.C.), BC294 from the American Cancer Society, and HD00700 from the United States Public Health Service (S.C.). S.C. is an American Cancer Society Research Professor.

Receptor Regulation
(*Receptors and Recognition*, Series B, Volume 13)
Edited by R. J. Lefkowitz
Published in 1981 by Chapman and Hall, 11 New Fetter Lane, London EC4P 4EE
© 1981 Chapman and Hall

3.1 INTRODUCTION

Since many aspects of the biology and chemistry of epidermal growth factors (EGF) have been reviewed recently (Carpenter and Cohen, 1978b, 1979; Carpenter, 1978), this manuscript will attempt to focus primarily on the interactions of EGF with membrane receptors. This has been for the past five years a rapidly evolving area of investigation by a number of groups. Although many gross facets of receptor control have been explored, little is actually known with certainty about the finer biochemical details. Within the limitations of this framework, we will attempt to integrate and evaluate present thinking concerning EGF:receptor inter-actions, focusing primarily on the work from the authors' laboratories.

3.2 CHEMISTRY OF EPIDERMAL GROWTH FACTORS

3.2.1 Mouse EGF

Mouse EGF (mEGF) is a single polypeptide chain of 53 amino acid residues (mol. wt. 6045), devoid of alanine, phenylalanine and lysine, that has an isoelectric point at pH 4.6 and an extinction coefficient ($E_{1cm}^{1\%}$ at 280 nm) of 30.9. The primary amino acid sequence and the location of the three intramolecular disulfide bonds are shown in Fig. 3.1.

The disulfide bonds in mEGF are required for biological activity. Extensive reduction of the disulfides in the presence of mercaptoethanol and urea yielded an inactive polypeptide. The biological activity, however, was completely restored by removal of the mercaptoethanol and urea by dialysis and subsequent reoxidation by air.

EGF lacking the carboxy-terminal Leu-Arg dipeptide also may be isolated from the submaxillary gland and is biologically active. Exposure of intact mEGF to mild tryptic digestion resulted in cleavage of the peptide bond between residues 49 and 53 and a derivative of mEGF composed of residues 1–48 that retained full biological activity when assayed *in vivo*, but only about 10 per cent of its activity when assayed with cultured fibroblasts. It is possible that when injected subcutaneously, EGF_{1-53} is converted into EGF_{1-48} which may be the circulating form of EGF *in vivo*. Alternatively, receptors on fibroblasts and epidermal cells may differ in their affinities for EGF and its derivatives.

Circular dichroic examination of EGF indicated the absence of significant α-helical structure and the presence of 22 per cent β-structure. It

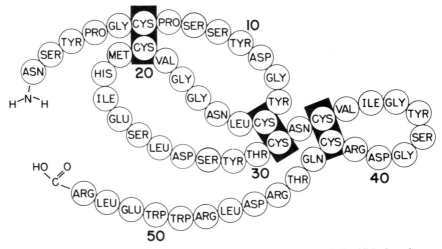

Fig. 3.1 Amino acid sequence of mEGF with placement of disulfide bonds
(from Savage *et al.*, 1973).

appears that mEGF exists as a very compact structure, due mainly to many
β-bends and the three disulfide bonds.

3.2.2 Human EGF

Since human fibroblasts possess receptors for mouse EGF, it was
hypothesized that a polypeptide homologous to mEGF existed in the
human. This was substantiated by the isolation of human EGF (hEGF)
from urine. Human EGF and mouse EGF appear to have identical biological properties; chemically, the two polypeptides are very similar, but not
identical. The completely unexpected finding by Gregory (1975) that a
relationship exists between mEGF and the human gastric antisecretory
hormone, urogastrone, was based on the comparison of the amino acid
sequences of the two polypeptides. Of the 53 amino acid residues compris-
ing each of the two polypeptides, 37 are common to both molecules, and
the three disulfide bonds are formed in the same relative positions.
Further, mEGF and urogastrone elicited nearly identical biological
responses; mEGF elicited a gastric antisecretory response, and urogastrone
produced the biological effects of EGF, as judged by its ability to induce
precocious eyelid opening in the newborn mouse. Thus far, urogastrone
and hEGF are the only naturally occurring polypeptides that compete with
[125]I-mEGF in radioreceptor assays. The available data strongly suggest that
human EGF and human urogastrone are identical. Additional information
on the chemistry of EGF can be located in the original references cited in

the previous review articles (Carpenter and Cohen, 1978b, 1979; Carpenter, 1978).

3.3 MECHANICS OF EGF:RECEPTOR INTERNALIZATION

3.3.1 Receptors for EGF

Specific, saturable receptors for EGF have been demonstrated using ^{125}I-labeled mEGF or hEGF (urogastrone) and a wide variety of cultured cells including corneal cells, human fibroblasts, lens cells, human glial cells, human epidermoid cells, 3T3 cells, granulosa cells, human vascular endothelial cells, human choriocarcinoma cells, and a number of other cell types. References may be found in Carpenter & Cohen (1979).

For different strains of human fibroblasts it has been estimated that each cell contains 40 000–100 000 binding sites for EGF, and apparent dissociation constants of $2–4 \times 10^{-10}$ M have been calculated (Carpenter *et al.*, 1975; Hollenberg and Cuatrecasas, 1975). The human epidermoid carcinoma cell line A-431 has the highest reported number of binding sites for EGF, $2–3 \times 10^6$ receptors per cell (Fabricant *et al.*, 1977; Haigler *et al.*, 1978).

Receptors for EGF were detected by O'Keefe *et al.* (1974) with crude membrane fractions prepared from a variety of mammalian tissues. These authors reported that placental and liver membranes have a high capacity to bind EGF. Specific binding of EGF has also been detected in liver membrane fractions of evolutionary distant organisms, such as certain teleosts and the dogfish shark (Naftel and Cohen, 1978). The specificity and high affinity of cells and membrane preparations for ^{125}I-EGF may be employed as the basis for competitive radioreceptor assays for EGF.

Indirect evidence suggests that the EGF receptor is a glycoprotein. Carpenter and Cohen (1977) demonstrated reversible inhibition of ^{125}I-EGF binding by a variety of lectins. Hock *et al.* (1979) reported that crosslinked EGF:receptor complexes are absorbed by immobilized lectins. Pratt and Pastan (1978) studied a mutant of 3T3 cells that has a lowered ^{125}I-EGF binding capacity and a decreased content of cell surface carbohydrate due to a block in the acetylation of glucosamine-6-phosphate. The binding was partially restored upon addition of *N*-acetylglucosamine to the media.

3.3.2 Internalization and degradation of EGF

It is now generally agreed that subsequent to the binding of EGF to the plasma membrane of the cell, the hormone, together with its receptor, enters the cell.

(a) *Biochemical evidence*

The first experimental data which indicated that EGF and its receptor were internalized resulted from the determination of the time-course of binding of ^{125}I-hEGF to human fibroblasts at 37° C and 0° C, shown in Fig. 3.2 (Carpenter and Cohen, 1976a). On continued incubation of the labeled hormone with fibroblasts at 37° C, the amount of cell-bound radioactivity decreased until a constant level of 15–20 per cent of the initial maximal amount of cell-bound radioactivity remained associated with the cells. When the cells were incubated with the labeled hormone at 0° C, there was no net loss of cell-bound radioactivity. Carpenter and Cohen (1976a) postulated that subsequent to the initial binding of ^{125}I-EGF to specific plasma membrane receptors, the EGF:receptor complex is internalized and the hormone is ultimately degraded in lysosomes. These conclusions were drawn from the following series of observations. 1. Cell-bound ^{125}I-EGF was rapidly degraded to mono [^{125}I]iodotyrosine at 37 °C. 2. At 0° C cell-

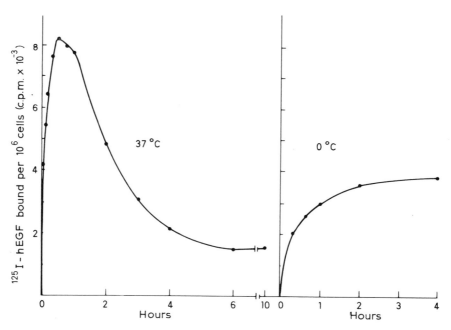

Fig. 3.2 Time-course of ^{125}I-hEGF binding to human fibroblasts at 37° C and 0° C. ^{125}I-hEGF (final concentration 4 ng/ml, 24 000 c.p.m./ng) was added to each culture dish containing the standard binding medium (1.5 ml for the 37° C experiments and 2.0 ml for the 0° C experiments). At the indicated time-intervals, duplicate dishes were selected and the cell-bound radioactivity was determined (from Carpenter and Cohen, 1976a).

bound ^{125}I-EGF was not degraded but slowly dissociated from the cell. 3. When the binding of ^{125}I-EGF was carried out at 37 ° C and the cells then incubated at 0 ° C, almost no release of cell-bound radioactivity was detected. 4. The degradation, but not the binding, required metabolic energy. 5. The degradation was inhibited by drugs that inhibit lysosomal function, such as chloroquine and ammonium chloride. 6. When ^{125}I-EGF was bound to cells at 0° C, the hormone was much more accessible to surface-reactive agents, such as trypsin and antibodies to EGF, than when the hormone was bound at 37 ° C. 7. Exposure of fibroblasts to EGF resulted in an apparent loss of plasma-membrane receptors for EGF, which suggested that the receptor also is internalized.

Internalization and degradation of ^{125}I-EGF have been confirmed in a number of laboratories with a variety of cells (granulosa cells, 3T3 and SV40-transformed 3T3 cells, human choriocarcinoma cells, epidermoid cells, and BSC-1 cells) using similar methodologies. The reader is referred to Carpenter and Cohen (1978b, 1979) for original references.

(b) Morphological evidence

The problem of direct visualization of the internalization of EGF has been approached in a number of laboratories using three procedures: the preparation and tracing of fluorescent derivatives of EGF, the tracing of ^{125}I-EGF by electron microscopy radioautography, and the preparation and tracing of EGF-ferritin conjugates by electron microscopy. With the exception of ^{125}I-EGF all of the other derivatives were covalent modifications of the single *N*-terminal amino group; all the derivatives retained substantial binding and biological activity.

Schlessinger *et al.* (1978b) and Shechter *et al.* (1978b) prepared a highly fluorescent rhodamine derivative of EGF and examined the binding and internalization of this derivative to 3T3 cells using image-intensified video fluorescent microscopy. Haigler *et al.* (1978) prepared fluorescein-conjugated EGF and examined the binding and internalization of this derivative by direct fluorescence microscopy using human epidermoid carcinoma cells (A-431) which are capable of binding much larger quantities of EGF than fibroblasts, thus rendering visualization possible. Both laboratories found that the initial binding of the derivatives to the cell surface was diffuse (except for a concentration of staining at the cell borders with the A-431 cells). Within 10–30 min at 37° C the labeled hormone was found within the cells in endocytotic vesicles. An intermediary patching stage was seen in the 3T3 cells, but was not detected in the A-431 cells; a 'microclustering' of receptors was suggested for the latter cells.

Gorden *et al.* (1978) used quantitative electron microscopic radioautography to localize ^{125}I-EGF in human fibroblasts. The initial binding of the labeled hormone was localized to the plasma membrane with some preference to coated pit regions. The membrane-bound ^{125}I-EGF was internalized by the cell in a time- and temperature-dependent fashion. The internalized grains were almost exclusively related to lysosomal structures.

Haigler *et al.* (1979a) have prepared a conjugate of epidermal growth factor and ferritin that retains substantial binding affinity for cell receptors and is biologically active. Glutaraldehyde-activated EGF was covalently linked to ferritin to produce a conjugate that contained EGF and ferritin in a 1:1 molar ratio. The conjugate was separated from free ferritin by affinity chromatography using antibodies to EGF. Monolayers of human epithelioid carcinoma cells (A-431) were incubated with ferritin-EGF (F-EGF) at 4° C and processed for transmission electron microscopy. Under these conditions, 6×10^5 molecules of F-EGF bound to the plasma membrane of each cell. In the presence of excess native EGF, the number of bound ferritin particles was reduced by 99 per cent, indicating that EGF:ferritin binds specifically to cellular EGF receptors. At 37° C, cell-bound F-EGF rapidly redistributed in the plane of the plasma membrane to form small groups that were subsequently internalized into pinocytic vesicles. By 2.5 min at 37° C, 32 per cent of the cell-bound F-EGF was localized in vesicles. After 2.5 min, there was a decrease in the proportion of conjugate in vesicles with a concomitant accumulation of F-EGF in multivesicular bodies. The authors suggest that the F-EGF-containing pinocytic vesicles evert upon fusion with multivesicular bodies. By 30 min, 84 per cent of the conjugate was located in structures morphologically identified as multivesicular bodies or lysosomes. These results are consistent with other morphological and biochemical studies utilizing ^{125}I-EGF and fluorescein-conjugated EGF.

3.3.3 The receptor for EGF and its internalization

Exposure of fibroblasts to EGF resulted in an apparent loss of plasma membrane receptors (down regulated) for EGF; the possibility was suggested (Carpenter and Cohen, 1976a) that receptors were internalized together with the hormone. It was suggested by Aharonov *et al.* (1978) that low concentrations of EGF result in the down regulation of unoccupied receptors. However, since the initial incubation with low concentrations of EGF were for an extended period of time (hours) and internalization occurs in minutes, it seems likely that new hormone–receptor complexes were formed and that more of the receptor was bound than expected. The question of EGF-induced internalization of unoccupied receptors is still open.

(a) Metabolism of crosslinked EGF:receptor complexes

Das, Fox and co-workers have provided information concerning the internalization and degradation of the receptor for EGF, utilizing a different approach. Das *et al.* (1977) and Das and Fox (1978) prepared a photoreactive derivative of ^{125}I-EGF and, following incubation of the derivative with 3T3 cells and photolysis, were able to detect in the cells a radioactive band on SDS-polyacrylamide gels with a molecular weight of 190 000, albeit in low yield (1.5–2 per cent). Incubation of affinity-labeled cells at 37° C resulted in the loss of the radioactive 190 000 mol. wt. band and the accumulation of three lower molecular weight radioactive bands (mol. wt. 62 000, 47 000 and 37 000). These presumed proteolytic products of the receptor were localized to the lysosomal fraction upon sub-cellular fractionation. Whether the covalent crosslinking of the receptor to EGF influenced its rate of degradation is not clear.

A novel interaction of ^{125}I-EGF (and possibly other ligands) with cell membranes has been reported by Baker *et al.* (1979) and Linsley *et al.* (1979). These workers have reported that a small fraction of cell-bound ^{125}I-EGF becomes covalently linked, by an unknown mechanism, to the presumed receptor, identified as 170 000–190 000 protein. Upon pro-longed incubation smaller fragments are noted similar to those observed in the studies with photoactivatable ^{125}I-EGF. The biological significance of this interesting direct-linkage phenomenon is not known.

A number of other attempts have been made to label the receptor for EGF. Wrann *et al.* (1979) and Wrann and Fox (1979) have reported the detection of the EGF receptor as a 170 000 mol. wt. protein on 3T3 cells and a 175 000 mol. wt. protein on A-431 cells by direct surface iodination. Sahyoun *et al.* (1978) covalently linked ^{125}I-labeled human EGF to rat liver membranes by a glutaraldehyde procedure and suggested a 'subunit' molecular weight of about 100 000 for the receptor. Hock *et al.* (1979) reported molecular weights of 160 000 and 180 000 for ^{125}I-labeled human EGF covalently cross-linked to human placental membranes.

(b) Morphological analysis of EGF:receptor topology

McKanna *et al.* (1979) have developed ultrastructural criteria for identifi-cation of the F-EGF:receptor complex, thereby enabling utilization of the F-EGF as an indirect marker to localize the receptor for this peptide hor-mone. The ferritin cores of bound F-EGF are situated 4–6 nm from the extracellular surface of the membrane. When cells were incubated for up to 30 min at 37° C, this characteristic spatial relationship was observed in all uptake stages (surface clustering, endocytosis and incorporation into multivesicular bodies), indicating that the hormone–receptor complex remains intact through these steps. However, when incubation was con-tinued for periods sufficient to allow hormone degradation (30–60 min), pools of free ferritin were observed in lysosomes. These authors also have developed a method for viewing the surface of intact cells *en face*, allowing closer scrutiny of the clustering of F-EGF:receptor complexes in the plane of the membrane prior to internalization. The particles in the F-EGF clusters observed by this method are spaced at 12 nm center-to-center, serving to set upper limits on the packing dimensions of the EGF:receptor complex.

(c) Mobility of EGF:receptor complexes

Employing fluorescent derivatives of EGF and insulin, Schlessinger et al. (1978a) were able to determine the lateral diffusion coefficients of the EGF–receptor and insulin–receptor complexes in 3T3 cells. Both hormone–receptor complexes were mobile on the cell surface, with similar diffusion coefficients, D (3–5) $\times 10^{-10}$ cm^2/s at 23° C, a diffusion coefficient much lower than that of a lipid probe. Increasing the temperature to 37° C resulted in rapid receptor immobilization. The immobilization was attributed to aggregation of hormone–receptor complexes or internalization.

In a very interesting experiment Maxfield et al. (1978) using fluorescent derivatives of EGF, insulin, and α_2-macroglobulin added to 3T3-4 cells, concluded that all three polypeptides were internalized within the same vesicles by a common pathway. It would appear that all of these ligand–receptor complexes migrate to specialized regions of the cell surface where pinocytosis occurs. The molecular mechanisms involved in the formation of these clusters and their subsequent pinocytosis are not understood. It has been suggested that these specialized regions are coated and form 'coated pits', although it is not at all certain that pinocytosis only occurs in coated regions. In fact the data of Haigler et al. (1979b) and Chinkers et al. (1979) show clearly that, in A-431 cells, EGF induces ruffling and that although most of the EGF enters the cell in small vesicles, at least some of the membrane-bound EGF enters the cell in large vesicles concerned with bulk pinocytosis.

To delineate the biological consequences of receptor clustering and internalization, it would be useful to have an agent which inhibits these processes. It has been reported (Maxfield et al., 1979a,b,c) that ammonia and amines block the clustering of EGF receptors on 3T3 cells when examined by their fluorescent enhancement methodology. However, Gorden et al. (1978) have reported that ^{125}I-EGF is found in lysosomes even in the presence of ammonia, and McKanna et al. (1979) have reported, using F-EGF and various amines, that internalization of the EGF–receptor complex into multivesicular bodies proceeded, but that hormone–receptor degradation was blocked. The differing reports on amine inhibition may be reconciled if one assumes that these agents slow the rate of internalization, but do not prevent internalization.

Fig. 3.3 diagramatically summarizes the authors' views on the mechanics of hormone–receptor internalization. Presumably this scheme would be modified by Maxfield and co-workers to include an amine inhibition of clustering of hormone–receptor complexes on the cell surface.

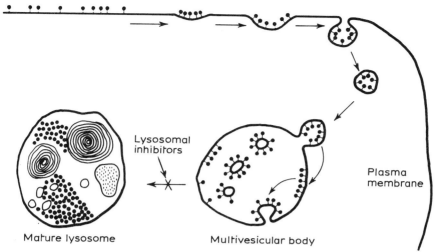

Fig. 3.3 Diagram of ferritin-EGF interactions with A-431 cells. Ferritin-EGF–receptor complexes, identified by the characteristic spatial relationship of particles and membranes (4 to 6 nm separation), are apparent at initial binding and are preserved through the processes of clustering, pinocytosis, and incorporation into MVBs. Further incubation at 37° C allows disruption of the ferritin-EGF–receptor complex (attested by pools of free ferritin), a process blocked by the presence of amines (from McKanna *et al.*, 1979).

3.3.4 Recovery of receptor activity

In the previous sections we have documented evidence that EGF–receptor complexes are rapidly and extensively removed from the cell surface by an active metabolic process. The results shown in Fig. 3.2 indicate that after incubation of human fibroblasts with ^{125}I-EGF for 6 h or longer a constant level of binding is maintained (Carpenter and Cohen, 1976a). This binding, representing approximately 25 per cent of the initial binding capacity, does not represent stable receptors that are not being internalized. Rather, this reduced level of binding reflects a steady-state in which the loss of hormone–receptor complexes from the cell surface is equalled by the insertion of unoccupied receptors into the plasma membrane (Carpenter and Cohen, 1976a; Haigler and Carpenter, 1980).

The apparent extent of down regulation, when examined in the continuous presence of labeled ligand as in Fig. 3.2, can be influenced by factors which increase or decrease either the rate of receptor loss or the rate of receptor replacement. That some cell lines only show a small decrease in binding with increasing periods of time may reflect a higher rate of receptor replacement.

Carpenter and Cohen (1976a) showed that following down regulation of 75 per cent of the ^{125}I-EGF-binding capacity of human fibroblasts, all of the original binding capacity could be recovered within 10 h by the addition of serum to ligand-free media. The serum-stimulated recovery process was blocked by cycloheximide or Actinomycin D. This suggests that either the synthesis of receptor molecules occurs during recovery or that the 'used' receptors are somehow recycled in a manner requiring macromolecular synthesis. Aharonov *et al.* (1978) have also observed the recovery process and reported that sparse, growing 3T3 cells recovered from down regulation much more quickly than did resting, confluent cells.

3.4 EGF-ENHANCED PHOSPHORYLATION AND THE EGF RECEPTOR

We have reviewed the evidence that EGF forms a complex with plasma membrane receptors, initiating an intricate series of biochemical and morphological events within the cell that includes internalization and degradation of the hormone–receptor complex. These events ultimately result in cell growth. It is reasonable to assume that the observed biochemical and morphological alterations induced by EGF result from the generation, amplification, and propagation of a series of 'signals'.

3.4.1 Characterization of EGF-enhanced phosphorylation

One approach to the problem of the nature of these signals is to develop a cell free system which responds in a measurable way to the presence of EGF. We have reported (Carpenter *et al.*, 1978, 1979) that membranes may be prepared from A-431 cells which retain the ability to bind ^{125}I-EGF *in vitro* and that following the formation of EGF–receptor complexes the capacity of these membranes to phosphorylate endogenous proteins in the presence of γ-labeled [^{32}P]ATP is enhanced. The net incorporation of ^{32}P into endogenous proteins was increased 2–3-fold by EGF. The A-431 membrane preparation appeared to have EGF-stimulated protein kinase activity toward exogenous substrates (histone) as well as toward membrane-associated proteins. These phosphorylation reactions did not depend on the presence of cyclic AMP or cyclic GMP. Both the endogenous phosphorylation and EGF-stimulated phosphorylation were dependent on the presence of Mg^{2+} or Mn^{2+}; Ca^{2+} was ineffective.

Partial acid hydrolysis and electrophoresis of the phosphorylated membranes showed that the major phosphorylated product, in both the presence and absence of EGF, was phosphotyrosine (Ushiro and Cohen, 1980). SDS-polyacrylamide gel electrophoresis and radioautography indicated that although EGF increased the phosphorylation of a number of membrane proteins, two components which appear to be glycoproteins with molecular weights of 170 000 and 150 000, were primarily affected (King *et al.*, 1980). A similar, but quantitatively much lower, phosphorylation effect of EGF was seen with membranes prepared from human fibroblasts or placenta (Carpenter *et al.*, 1980).

The activation of the kinase by EGF appears to be a reversible phenomenon since removal of EGF by anti-EGF IgG results in a 'deactivation' of the kinase to the original basal level of activity (Cohen *et al.*, 1980).

3.4.2 Purification of the EGF–receptor–kinase complex

The membrane preparation may be solubilized by a number of non-ionic detergents with the retention of both ^{125}I-labeled EGF binding activity (Carpenter, 1979b) and EGF-enhanced phosphorylation of the specific membrane proteins (Cohen *et al.*, 1980). Gel filtration of the Triton X-100-solubilized membrane preparation on Sephacryl S-300 indicated that most of the ^{125}I-EGF-binding activity and EGF-stimulated phosphorylation activity elute together in a volume which suggests a molecular size of about 300 000–4 000 000. It is not known whether or not the kinase activity and/or the receptor are present as Triton micelles. If both are indeed in micelles, the efficiency of the EGF-stimulated phosphorylation reaction suggests they are in the same micelle and were probably, therefore, associated together in the intact membrane.

In an attempt to understand the interactions leading to EGF-enhanced kinase activity, we have purified the receptor by affinity chromatography using EGF covalently linked to Affi-Gel. In brief, Triton extracts of A-431 membranes were stirred with EGF-Affi-Gel and the gel was washed exhaustively. Elution was successfully carried out by the addition of 5 mM-ethanolamine (or by the addition of a large excess of EGF).

The ethanolamine eluate possessed the capacity not only to bind ^{125}I-EGF but also to respond to the addition of EGF by an enhanced phosphorylation of endogenous substrates.

We have compared: (a) the composition of the original solubilized membrane preparation with the material eluted from the EGF-Affi-Gel (by SDS–polyacrylamide gel electrophoresis and Coomassie blue staining) and (b) the nature of the components phosphorylated in the original extract with those phosphorylated in the material eluted from the EGF-Affi-Gel (by SDS–polyacrylamide gel electrophoresis and radioautography). The elutions were carried out using either the EGF procedure or the ethanolamine procedure.

We conclude from the Coomassie blue data shown in Fig. 3.4 that (a) the composition of the non-adsorbed material in the Triton extracts (lane

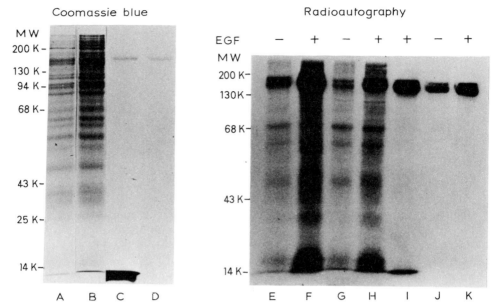

Fig. 3.4 Electrophoresis and autoradiography of solubilized A-431 membranes purified by affinity chromatography. A-431 membranes (10 mg protein) were solubilized in 800 μl of 20 mM-Hepes buffer, pH 7.4, containing 1 per cent Triton X-100 and 10 per cent glycerol. After centrifugation, aliquots (200–400 μl) were adsorbed to EGF-Affi-Gel beads and the beads were washed. The adsorbed material was eluted by either EGF or ethanolamine. Aliquots of the original extract, the non-adsorbed material, and the EGF or ethanolamine eluates were subjected to the standard phosphorylation procedures in the presence and absence of EGF for 10 min at 0° C. The reactions were stopped and analyzed by SDS gel electrophoresis, Coomassie blue staining, and radioautography. A–D: Coomassie blue stain of original extract (A), non-adsorbed material (B), EGF eluate (C), and ethanolamine eluate (D). E–K: radioautography of membrane components phosphorylated in absence (E, G, J) and presence (F, H, I, K) of EGF. E and F, original extract; G and H, non-adsorbed material; I, EGF eluate; J and K, ethanolamine eluate (from Cohen *et al.* 1980).

B) is almost identical to that of the original extract (lane A); (b) elution of the gel with either EGF or ethanolamine results in identical patterns (lanes C and D) consisting of one major 150 K band. Trace quantities of other bands (not apparent in the photograph) are detectable on the original gels at the 170 K, 130 K and 50–60 K regions. Thus, we have achieved by this method a very considerable purification.

We conclude from the radioautography data shown in Fig. 3.4 (lanes E–K) that under the conditions employed the major phosphorylated component in the original Triton extract was a doublet in the 150–170 K region (lanes E and F). In the non-adsorbed Triton supernate, the patterns were similar except that the EGF-stimulated 150–170 K components had decreased (lanes G and H). In both the EGF and ethanolamine eluates, only this doublet was detected, with the major radioactive band (150 K) corresponding to the major Coomassie blue-staining material. The stimulatory effect of adding EGF to the ethanolamine eluate during the phosphorylation is clearly seen (compare lanes J and K).

To summarize, the ethanolamine eluate, which by SDS electrophoresis contains one major protein band and a few trace bands, possesses both ^{125}I-EGF binding capacity and EGF-stimulated phosphorylation activity. The major 150 K protein band is a substrate of the phosphorylation reaction. The evidence suggests that the major 150 000 protein band is the receptor for EGF and that it is also the major substrate for the EGF-induced phosphorylation reaction.

The adsorption to the affinity gel of EGF-binding activity together with EGF-stimulated phosphorylation activity suggests an inherent association of these two activities, since only the receptor could be assumed to have an affinity for the EGF-Affi-Gel. We have assumed, based on data (Carpenter *et al.*, 1979) which clearly indicate different heat sensitivities of the receptor and the kinase, that two separate entities are involved and that one of the trace bands represents the protein kinase. It is conceivable that the receptor and kinase activities are present in the same molecule.

3.4.3 Significance

The biological role of the EGF-enhanced phosphorylation reaction is not known and one can only speculate concerning its possible relationship to nutrient transport, hormone–receptor internalization, modification of cytoskeletal elements, mitogenic signalling or some other aspect of the 'pleiotropic' effect of EGF.

Insulin and EGF have been reported to stimulate the phosphorylation of the S6 ribosomal protein in 3T3-L1 preadipocyte (Smith *et al.*, 1979). It has been suggested that the phosphorylation of this protein is involved in the transition of quiescent cells to growing cells. EGF has been reported to increase the phosphorylation of nuclear f_1 histone in chick epidermis (Huff and Guroff, 1978). The phosphorylation of a 33 000 mol. wt. membrane protein in 3T3 cells in the presence of another growth factor, FGF, has also been noted (Nilsen-Hamilton and Hamilton, 1979).

3.5 RELATIONSHIP OF BINDING TO BIOLOGICAL ACTIVITY

In the previous section we have reviewed what is known about the biochemical mechanics by which EGF interacts with cell surface receptors and is subsequently internalized and degraded. An important question is how events involved in hormone recognition and metabolism may be related to biological responses. Various approaches to this problem have been employed, but to date convincing answers have not been forthcoming. Nevertheless, some points are reasonably clear and can form a conceptual framework to discuss this important topic.

3.5.1 Rapid changes in cell physiology

Clearly the biological responses produced by EGF are numerous and several separate, but interrelated, mechanisms may be involved. The most rapid effects observed upon the addition of EGF to cultured cells are morphological changes (Chinkers *et al.*, 1979) and increased rates of transport (see Carpenter and Cohen, 1979 for review). Using scanning electron microscopy, Chinkers *et al.* (1979) have demonstrated that within five minutes of exposure of A-431 tumor cells to EGF a large increase in membrane ruffling and extension of filopodia is apparent. These changes can be observed at very short intervals (60 s) after addition of EGF and are transient in nature, subsiding within 15 min. Haigler *et al.* (1979b) measured the effect of EGF on the uptake of horseradish peroxide (HRP) in A-431 cells and demonstrated a large (10-fold) increase in the rate of fluid-phase pinocytosis in temporal coordination with enhanced ruffling. The ability of EGF to increase the rate of HRP uptake was detectable 30 s after addition of the growth factor and was transient in nature, returning to control levels within 15 min. Similar events probably occur but to a less marked extent in cells having lower concentrations of receptors. Since these uptake and morphological alterations occur very rapidly, degradation of cell-bound EGF is not likely to be necessary. Since increased HRP uptake occurs very rapidly and requires the occupancy of only a small fraction of available receptors (4 percent occupancy for half-maximal uptake), the signal for this cellular response to EGF occurs during clustering or initial internalization of hormone–receptor complexes. A candidate for such a signal is the EGF-sensitive protein kinase activity present in cell membranes (Carpenter *et al.*, 1978, 1979).

Similar analyses probably apply to the increases in active transport of ions and low-molecular-weight nutrients (reviewed in Carpenter and Cohen, 1979).

3.5.2 Stimulation of DNA synthesis

The relationship of responses such as increased DNA synthesis to binding, internalization and degradation is more difficult to assess. The primary complication is that enhanced DNA synthesis is observed no sooner than 12 h after the addition of EGF.

Two points relating EGF binding to maximal stimulation of DNA synthesis are reasonably clear: (1) persistent interactions between EGF and cell-surface receptors must occur for many hours and (2) occupancy of approximately 25 per cent of the available binding sites is necessary. The occupancy ratio has been reported by several different investigators (Hollenberg and Cuatrecasas, 1975; Carpenter and Cohen, 1976b; Aharonov *et al*., 1978). However, the interpretation of this fractional occupancy ratio is difficult as binding analyses are complicated and the possible functional heterogeneity of binding sites has not been addressed.

To investigate the length of time cells needed to be exposed to EGF before they became committed to increased DNA synthesis, Carpenter and Cohen (1976b) added antibody to EGF to cultures of human fibroblasts at various times after the addition of EGF. They then related the length of EGF exposure to the capacity of the cells to increase DNA synthesis. Removal of EGF from the media by antibody addition of 0–6 h prevented any stimulation of DNA synthesis and not until 12 h was the level of stimulation of DNA synthesis refractory to the antibody. Similar results have been published by Lindgren and Westermark (1976, 1977), Aharonov *et al*. (1978) and Shechter *et al*. (1978a). Haigler and Carpenter (1980) also reported similar results using either antibody to EGF or antibody to the EGF receptor to interrupt hormone–receptor interactions.

Shechter *et al*. (1978a) have reported that although the addition of anti-EGF antibody to cultured fibroblasts at times up to 8 h after the addition of EGF blocks enhanced DNA synthesis, removal of the growth factor by washing 30 min after its addition does not completely prevent stimulation of DNA synthesis. They propose that a small fraction of the cell-bound EGF is not removed by washing and does not dissociate into EGF-free media for at least 8 h. The authors' data further indicate that this very tightly bound EGF remains on the extracellular surface of the cells, since it can be inactivated by antibodies to EGF. Shechter and his colleagues suggest that this very high-affinity binding of EGF represents a negligible fraction of the total binding and is necessary for the induction of DNA synthesis. The covalent linkage of EGF to its receptor (Baker *et al*., 1979; Linsley *et al*., 1979) might explain the necessary high-affinity binding to support these observations. However, chemical modification of the amino terminus of EGF prevents covalent cross-linking to the receptor, but does not reduce the binding activity. The conclusion of Shechter *et al*. (1978a) would also require that the high-affinity EGF–receptor complexes are not subject to internalization and degradation. Such a subclass of receptors has not been detected. It should be noted that in experiments reported by other groups (Lindgren and Westermark, 1976, 1977; Aharonov *et al*., 1978) removal of EGF from cultured cells by washing did prevent stimulation of DNA synthesis.

In a subsequent publication Shechter *et al*. (1979) have addressed the question of whether the clustering or aggregation of EGF–receptor com-

plexes is required for the stimulation of DNA synthesis. Cultures of fibroblasts were incubated with either biologically inactive cyanogen bromide-
treated EGF (which has a 10-fold lower binding affinity compared with native
EGF) or with a concentration of native EGF too low to be biologically effective (0.1 pg/ml). Neither of these additions stimulated DNA synthesis; however, addition of small amounts of anti-EGF antibodies shortly (20 min) after
either the cyanogen bromide-treated EGF or the low amounts of native EGF
produced a stimulation of DNA synthesis that was 50–100 per cent of the
maximal stimulation obtained by native EGF at optimal concentrations.
Monovalent Fab' antibody was not effective and no enhancement of DNA
synthesis was produced by the divalent antibody alone. The authors suggested
that the divalent antibody evoked EGF activity by cross-linking and aggrega-
ting otherwise inactive hormone–receptor complexes. Although only a very
restricted range of antibody dilutions were effective, the authors indicate that
the local aggregation of EGF–receptor complexes is a necessary step for the
induction of DNA synthesis. In similar experiments Kahn *et al.* (1978)
demonstrated that insulin–receptor complexes must aggregate to produce the
appropriate biological response.

Maxfield *et al.* (1979a,b,c) have reported that bacitracin and certain amines
(ammonium salts, ethylamine, propylamine, *n*-butylamine) block the clustering of EGF–receptor complexes on the cell surface and thereby prevent internalization. This group (Maxfield *et al.*, 1979a) reports that these putative
inhibitors of clustering do not block stimulation of DNA synthesis by EGF,
but actually potentiate the effect of the growth factor. It should be noted that
this effect of amines on the EGF stimulation of DNA synthesis was only
observed in very restricted conditions, a limited (2.5 h) exposure of quiescent
cells to EGF in the presence or absence of the amines. This class of amines
has been reported to inhibit clustering and internalization of insulin–receptor
complexes in hepatocytes without blocking the capacity of insulin to stimulate
amino acid transport (LeCam *et al.*, 1979). The evidence to support the
capacity of these inhibitors to block clustering and internalization, however, is
derived solely by morphological criteria, i.e. by using an image-intensifier
television camera to view the interaction of fluorescein or rhodamine-
conjugated ligands with intact cells (Willingham and Pastan, 1978). Although
this technology certainly has advantages, resolution in terms of dimensions of
the aggregates and their specific location would appear to be difficult.

Using ferritin-conjugated EGF and electron microscopy, McKanna *et al.*
(1979) demonstrated that EGF–receptor complexes are internalized in the
presence of amines. Also, Gorden *et al.* (1978) performed quantitative
electron-microscopic radioautography after incubating cultured fibroblasts
with ^{125}I-labeled EGF in the presence and absence of ammonium chloride.
Their data also indicate that hormone–receptor complexes are internalized in
the presence of this inhibitor. Thus, conclusions regarding the relationship of

the internalization of EGF–receptor complexes to enhanced DNA synthesis which are based solely on amine inhibition are not yet warranted.

In summary the available data show that binding of EGF to specific cell surface receptors is obligatory for stimulation of DNA synthesis and that clustering of hormone–receptor complexes is involved. However, the possible role of internalization and degradation in the production of biological responses by EGF is not resolved.

Fox and Das (1979) and Heldin *et al.* (1979) have presented indirect evidence to support the idea that internalization is an important step in the activation of DNA synthesis by EGF. Other groups (Aharonov *et al.*, 1978; Schechter *et al.*, 1979; Maxfield *et al.*, 1979a) conclude that internalization is not necessary for stimulation of DNA synthesis and actually may be a mechanism for stopping the generation of intracellular mitogenic signals at the cell surface.

3.6 OTHER CONTROLS OF RECEPTOR ACTIVITY

3.6.1 Transforming agents

Several studies have shown interesting relationships between effects of EGF and the transformed state or between EGF and transforming agents. Carpenter and Cohen (1976b, 1978a) demonstrated that human fibroblasts grown in the continuous presence of EGF did not exhibit two growth-controlling mechanisms otherwise associated with the regulated proliferation of these cells. When grown in the continuous presence of EGF, the cells were not restricted to a tightly packed confluent monolayer but rather formed populations several layers thick. Also, the growth of cells in media containing a low concentration of serum or deficient sera was not impeded when EGF was present. It was noted that these behaviors of cells *in vitro* were more similar to those of transformed cells than normal cells. A notable exception was the inability of EGF to promote anchorage-independent growth, which is the characteristic of cultured cells most closely associated with malignant potential.

Studies of ^{125}I-EGF binding to cultured cells by Carpenter *et al.* (1975) showed that normal rat kidney cells bound the growth factor, but the same cells transformed by the Kirsten sarcoma virus did not. They noted that transformation by DNA viruses did not reduce binding capacity and that several other cell lines which were reportedly infected with RNA-transforming viruses exhibited low ^{125}I-EGF binding capacity.

Studies of ^{125}I-EGF binding to pairs of normal and transformed cell lines were reported by Todaro *et al.* (1976, 1977). Their results showed that transformation by murine or feline sarcoma viruses consistently decreased ^{125}I-

EGF binding capacity in different lines of normal cells. In these studies the activity of receptors for other ligands, such as multiplication-stimulating activity, was not affected by transformation. Also, cells transformed by DNA viruses, chemicals (with a few exceptions), and spontaneous transformants did not exhibit altered ^{125}I-EGF binding capacity. These authors postulated several possible explanations for the selective reduction of ^{125}I-EGF binding capacity by sarcoma viruses. One explanation was that sarcoma transformation resulted in the production of an EGF-like molecule which would lower the apparent binding capacity for exogenous ^{125}I-EGF by direct competition. Investigation of this possibility has revealed that sarcoma transformed cell lines excrete a polypeptide termed sarcoma growth factor (SGF) which is capable of stimulating cell growth and which apparently interacts with the EGF receptor (DeLarco and Todaro, 1978a; Todaro and DeLarco, 1978; Todaro *et al.*, 1979). *In vitro* SGF not only stimulates cellular proliferation but is able to promote the anchorage-independent growth of non-transformed cells, an effect not produced by EGF. SGF competes with ^{125}I-EGF in specific radioreceptor assays, but is immunologically distinct from EGF. Guinivan and Ladda (1979) report the presence of a factor in the culture media of Kirsten sarcoma transformed cells which stimulates DNA synthesis in normal cells, but enhances rather than inhibits ^{125}I-EGF binding. Pruss *et al.* (1978) have demonstrated that 3T3 variants lacking the EGF receptor are readily transformed by the Kirsten sarcoma virus. This result suggests that either interaction of SGF with the EGF receptor is not necessary for transformation or that the sarcoma factor interacts with sites other than the EGF receptor. It will be important to determine the chemical nature of SGF and if this growth factor binds to cellular sites other than the EGF receptor. Experiments concerning the possible activities of SGF *in vivo* have not been reported. It would be helpful to know whether SGF mimics EGF in the newborn mouse eyelid-opening assay which is the most specific index of EGF biological activity.

These results have provided an impetus to search for the ectopic production of other growth factors by transformed cells. Sherwin *et al.* (1979) have detected the production of nerve growth factor by melanoma cells and DeLarco and Todaro (1978b) have reported the formation of multiplication stimulating activity by fibrosarcoma cells. Todaro and his colleagues (1979) have suggested that the ectopic production of growth factors by cells that also respond to the same factor(s) may offer a selective advantage to such cells and provide means by which some transformation properties are acquired. A reduction in EGF–receptor activity has been noted in some instances of chemical transformation, particularly by benzopyrene (Todaro *et al.*, 1976; Hollenberg *et al.*, 1979; Brown *et al.*, 1979b). However, no evidence has been reported for the ectopic production by these cells of factors able to interact with the EGF-receptor.

3.6.2 Tumor promoters

EGF has been shown to act as tumor promoter *in vivo* in studies involving the application of methylcholanthrene to mouse skin (see Carpenter and Cohen, 1979 for review). Another potent tumor promoter is 12-*O*-tetradecanoyl phorbol-13-acetate (TPA) which is derived from croton oil. In cell culture systems it has been demonstrated that TPA at concentrations of 10^{-8} to 10^{-10} M blocks ^{125}I-EGF binding to cell surface receptors (Lee and Weinstein, 1978, 1979; Shoyab *et al.*, 1979; Brown *et al.*, 1979a; Murray and Fusenig, 1979). Derivatives of TPA which are inactive promoters *in vivo* do not block ^{125}I-EGF binding. The mechanism by which TPA affects ^{125}I-EGF binding is not clear, but appears to be indirect, i.e. TPA does not compete for ^{125}I-EGF binding in the same manner as unlabeled EGF. At present the indirect effect of TPA is thought to result from perturbations of the phospholipid microenvironment near the EGF receptor or by TPA binding to the receptor at a site other than the EGF binding site. However, even these mechanisms are likely to be simplifications as no effect of TPA on ^{125}I-EGF binding in membrane preparations has been demonstrated. Studies with cultured cells indicate that the TPA effect on EGF binding (1) is specific (binding of insulin, multiplication stimulating activity, concanavalin A, nerve growth factor, low density lipoprotein, or murine type C ectopic viral glycoprotein [gp60] is not altered), (2) is strongly temperature dependent, (3) is energy dependent, and (4) results in a lowered receptor affinity for EGF without a decrease in receptor number.

EGF and TPA produce a number of similar responses in cultured cells (Lee and Weinstein, 1978, 1979) and TPA acts synergistically with EGF (and other mitogens) to enhance the stimulation of DNA synthesis (Dicker and Rozengurt, 1979; Frantz *et al.*, 1979). In regard to the potentiation of biologic responses to mitogenic agents the action of TPA is not specific for EGF, but seems to produce an enhancement of mitogenic activities in general. TPA also produces mitogenic activities in cells that do not respond to EGF.

3.6.3 Differentiation

In what may be an important area of receptor regulation several reports have examined the influence of cell differentiation on ^{125}I-EGF binding. Vlodavsky *et al.* (1978) report that cultured bovine granulosa cells respond to EGF by increasing cell numbers, and have approximately 23 000 EGF receptors per cell. When these cells spontaneously differentiate in culture to luteal cells, they no longer respond to EGF. Surprisingly, the loss of sensitivity to EGF is accompanied by a 5-fold increase in the number of receptors per cell (105 000 receptors per cell). Both granulosa and luteal cells internalize and degrade ^{125}I-EGF and the receptor K_D was only slightly altered ($K_D = 2.4 \times 10^{-10}$ M for granulosa cells versus $K_D = 7.0 \times 10^{-10}$ M for luteal cells).

Huff and Guroff (1979) have studied the rat pheochromocytoma clone PC12 as a model of neuronal development. This cell line has specific high-affinity EGF receptors and ornithine decarboxylase activity is increased by the addition of EGF. When the PC12 cell line is grown in the presence of nerve growth factor for several days many morphological, biochemical and electrical aspects of neuronal differentiation appear. At this time the binding capacity for ^{125}I-EGF is decreased by approximately 80 per cent and ability of the EGF to increase ornithine decarboxylase activity is similarly reduced.

Rees *et al.* (1979) have examined the differentiation of cultured mouse teratocarcinoma stem cells. These undifferentiated embryonal carcinoma cells are pleuripotent and highly tumorgenic. When grown *in vitro* for several days the embryonal carcinoma cells form morphologically distinct endoderm-like cells that have a reduced tumorgenic potential. ^{125}I-EGF binding is barely detectable in the embryonal carcinoma cells, but increases approximately 12-fold as the cells develop into endoderm-like cells. At this time the apparently differentiated cells are sensitive to the mitogenic activity of EGF.

3.6.4 Miscellaneous

(a) Lectins and glycoprotein metabolism
The activity of the EGF receptor can be inhibited by interfering with glyco-protein synthesis or by agents that react with glycoprotein. Carpenter and Cohen (1977) demonstrated that lectins block ^{125}I-EGF binding at 37° C or at 0° C and in a reversible manner. Pratt and Pastan (1978) studied ^{125}I-EGF binding in a 3T3 mutant cell line defective in glycoprotein synthesis due to a partial block in the acetylation of glucosamine-6-phosphate. This block in carbohydrate processing could be overcome by exogenous N-acetylglucosamine in the media. In the absence of added N-acetylglucosamine, ^{125}I-EGF binding was reduced approximately 80 per cent in the mutant. Feeding N-acetylglucosamine to the mutant for several days partially restored the ^{125}I-EGF binding capacity.

(b) Glucocorticoids
EGF binding capacity can be affected by incubation of cells for several days in media supplemented with glucocorticoids (Baker *et al.*, 1978; Baker and Cunningham, 1978). Cultured human fibroblasts grown for several days in media containing 100 ng of dexamethasone/ml exhibited a 50–100 per cent increase in ^{125}I-EGF binding capacity and an increased responsiveness to the mitogenic action of EGF.

(c) Modulation of protein synthesis

Modulation of protein synthesis also affords an experimental technique to influence EGF receptor activity. Carpenter (1979a) reported a half-life of approximately 15 h for EGF receptor activity when protein synthesis was inhibited in human fibroblasts. Aharonov *et al.* (1978), using cycloheximide to inhibit protein synthesis, calculated a similar value ($t_{1/2}$) for the EGF receptor in 3T3 cells. In the study carried out by Carpenter (1979a) protein synthesis was stopped by the removal of histidine from the media and the addition of L-histidinol to create a stringent amino acid starvation. In those experiments protein synthesis could be rapidly reinitiated by the addition of L-histidine. When protein synthesis was reinitiated by this procedure EGF receptor activity was completely recovered within 10 h and the recovery process was stimulated by the presence of fresh serum. Also, complete recovery took place in the presence of Actinomycin D.

It should be noted that culture conditions can influence the capacity of cells to bind ^{125}I-EGF. Such variables as serum concentration (Carpenter, unpublished results), cell density (Pratt and Pastan, 1978; Brown *et al.*, 1979; Bhargava, 1979), and the age of the culture (Bhargava, 1979) have been noted to influence growth-factor binding.

3.7 SUMMARY

The last few years have seen remarkable advances in our understanding of the mechanics by which EGF interacts with cell surfaces and the processes by which the cell-bound growth factor is sequestered and metabolized. Studies are beginning to relate the steps in hormone–receptor formation, aggregation, internalization, and degradation to the production of biological responses. Also, a system has been described in which the formation of EGF–receptor complexes *in vitro* produces a biochemical response *in vitro*. This may offer a means of dissecting at a molecular level individual components involved in generating at least part of the pleiotropic response of cells to mitogens. As often is the case in scientific progress advances in technology have been a key. In this instance the widespread use of cell culture technologies and the ability to radiolabel polypeptides to high specific radioactivities have been crucial.

Unfortunately, recent progress in understanding the mechanism of action of EGF at the cellular and subcellular levels has not been matched at the more complex level of the intact organism. It is extremely important that we should determine what physiological function(s) endogenous EGF has in mammalian biology. We are hopeful that information will become available to answer this question.

REFERENCES

Aharonov, A., Pruss, R.M. and Herschman, H.R. (1978), *J. Biol. Chem.*, **253**, 3970–3977.

Baker, J.B. and Cunningham, D.D. (1978), *J. Supramol. Struct.*, **9**, 69–77.

Baker, J.B., Barsh, G.S., Carney, D.H. and Cunningham, D.D. (1978), *Proc. Natl. Acad. Sci. USA*, **75**, 1882–1886.

Baker, J.B., Simmer, R.L., Glenn, K.C. and Cunninghan, D.D. (1979), *Nature*, **278** 743–745.

Bhargava, G., Rifas, L. and Makman, M.H. (1979), *J. Cell Physiol.*, **100**, 365–374.

Brown, K.D. and Holley, R.W. (1979), *J. Cell Physiol.*, **100**, 139–146.

Brown, K.D., Dicker, P. and Rozengurt, E. (1979a), *Biochem. Biophys. Res. Commun.*, **86**, 1037–1043.

Brown, K.D., Yeh, Y.C. and Holley, R.W. (1979b), *J. Cell Physiol.*, **100**, 227–238.

Carpenter, G. (1978), *J. Invest. Dermatol.*, **71**, 283–287.

Carpenter, G. (1979a), *J. Cell Physiol.*, **99**, 101–106.

Carpenter, G. (1979b), *Life Sci.*, **24**, 1691–1698.

Carpenter, G. and Cohen, S. (1976a), *J. Cell Biol.*, **71**, 159–171.

Carpenter, G. and Cohen, S. (1976b), *J. Cell Physiol.*, **88**, 227–237.

Carpenter, G. and Cohen, S. (1977), *Biochem. Biophys. Res. Commun.*, **79**, 545–552.

Carpenter, G. and Cohen, S. (1978a), in *Molecular Control of Proliferation and Differentiation'* (J. Papaconstantinou, ed.), Academic Press, New York, pp. 13–31.

Carpenter, G. and Cohen, S. (1978b), in *Biochemical Actions of Hormones*, Vol. 6 (G. Litwack, ed.), Academic Press, New York, pp. 203–247.

Carpenter, G. and Cohen, S. (1979), *Annu. Rev. Biochem.*, **48**, 193–216.

Carpenter, G., King, L., Jr. and Cohen, S. (1978), *Nature*, **276**, 409–410.

Carpenter, G., King, L., Jr. and Cohen, S. (1979), *J. Biol. Chem.*, **254**, 4884–4891.

Carpenter, G., Poliner, L. and King, L. (1980), Molec. and Cellular Endocrinol, **18**, 189–199.

Carpenter, G., Lembach, K.J., Morrison, M.M. and Cohen, S. (1975), *J. Biol. Chem.*, **250**, 4297–4304.

Chinkers, M., McKanna, J.A. and Cohen, S. (1979), *J. Cell Biol.*, **83**, 260–265.

Cohen, S., Carpenter, G. and King, L., Jr. (1980), *J. Biol. Chem.*, **255**, 4834–4842.

Das, M. and Fox, C.F. (1978), *Proc. Natl. Acad. Sci. USA*, **75**, 2644–2648.

Das, M., Miyakawa, T., Fox, C.F., Pruss, R.M., Aharonov, A. and Herschman, H.R. (1977), *Proc. Natl. Acad. Sci. USA*, **74**, 2790–2794.

DeLarco, J.E. and Todaro, G.J. (1978a), *Proc. Natl. Acad. Sci. USA*, **75**, 4001–4005.

DeLarco, J.E. and Todaro, G.J. (1978b), *Nature*, **272**, 356–358.

Dicker, P. and Rozengurt, E. (1979), *Nature*, **276**, 723–726.

Fabricant, R.N., DeLarco, J.E. and Todaro, G.J. (1977), *Proc. Natl. Acad. Sci. USA*, **74**, 565–569.

Fox, C.F. and Das, M. (1979), *J. Supramol. Struct.*, **10**, 199–214.

Frantz, C.N., Stiles, C.D. and Scher, C.D. (1979), *J. Cell Physiol.*, **100**, 413–424.

Gorden, P., Carpentier, J.L., Cohen, S. and Orci, L. (1978), *Proc. Natl. Acad. Sci. USA*, **75**, 5025–5029.

Gregory, H. (1975), *Nature*, **257**, 325–327.

Guinivan, P. and Ladda, R.L. (1979), *Proc. Natl. Acad. Sci. USA*, **76**, 3377–3381.

Haigler, H.T., Ash, J.F., Singer, S.J. and Cohen, S. (1978), *Proc. Natl. Acad. Sci. USA,* **75**, 3317–3321.

Haigler, H.T. and Carpenter, G. (1980), *Biochim. Biophys. Acta.,* **598**, 314–325.

Haigler, H.T., McKanna, J.A. and Cohen, S. (1979a), *J. Cell Biol.,* **81**, 382–395.

Haigler, H.T., McKanna, J.A. and Cohen, S. (1979b), *J. Cell Biol.,* **83**, 82–90.

Heldin, C.H., Westermark, B. and Wasteson, Å. (1979), *Nature,* **282**, 419–420.

Hock, R.A., Nexo, E. and Hollenberg, M.D. (1979), *Nature,* **277**, 403–405.

Hollenberg, M.D., Barrett, J.C., Ts'o, P.O.P. and Berhanu, P. (1979), *Cancer Res.,* **39**, 4166–4169.

Hollenberg, M.D. and Cuatrecasas, P. (1975), *J. Biol. Chem.,* **250**, 3845–3853.

Huff, K.R. and Guroff, G. (1978), *Biochem. Biophys. Res. Commun.,* **85**, 464–472.

Huff, K.R. and Guroff, G. (1979), *Biochem. Biophys. Res. Commun.,* **89**, 175–180.

Kahn, C.R., Baird, K.L., Jarrett, D.B. and Flier, J.S. (1978), *Proc. Natl. Acad. Sci. USA,* **75**, 4209–4213.

King, L., Jr., Carpenter, G. and Cohen, S. (1980), *Biochemistry,* **19**, 1524–1528.

LeCam, A., Maxfield, F., Willingham, M. and Pastan, I. (1979), *Biochem. Biophys. Res. Commun.,* **88**, 873–881.

Lee, L.S. and Weinstein, I.B. (1978), *Science,* **202**, 313–315.

Lee, L.S. and Weinstein, I.B. (1979), *Proc. Natl. Acad. Sci. USA,* **76**, 5168–5172.

Lindgren, A. and Westermark, B. (1976), *Exp. Cell Res.,* **99**, 357–362.

Lindgren, A. and Westermark, B. (1977), *Exp. Cell Res.,* **106**, 89–93.

Linsley, P.S., Blifeld, C., Wrann, M. and Fox, C.F. (1979), *Nature,* **278**, 745–748.

Maxfield, F.R., Davies, P.J.A., Klempner, L., Willingham, M. and Pastan, I. (1979a), *Proc. Natl. Acad. Sci. USA,* **76**, 5731–5735.

Maxfield, F.R., Schlessinger, J., Shechter, Y., Pastan, I. and Willingham, M.C. (1978), *Cell,* **14**, 805–810.

Maxfield, F.R., Willingham, M.C., Davies, P.J.A. and Pastan, I. (1979b), *Nature,* **277**, 661–663.

Maxfield, F.R., Willingham, M.C., Schlessinger, J., Davies, P.J.A. and Pastan, I. (1979c), *Cold Spring Harbor Symp. Cell Prolif.,* **6**, 159–166.

McKanna, J.A., Haigler, H.T. and Cohen, S. (1979), *Proc. Natl. Acad. Sci. USA,* **76**, 5689–5693.

Murray, A.W. and Fusenig, N.E. (1979), *Cancer Lett.,* **7**, 71–77.

Naftel, J. and Cohen, S. (1978), *J. SC Med. Assoc.,* **74**, 53.

Nilsen-Hamilton, M. and Hamilton, R.T. (1979), *Nature,* **279**, 444–446.

O'Keefe, E., Hollenberg, M.D. and Cuatrecasas, P. (1974), *Arch. Biochem. Biophys.,* **164**, 518–526.

Pratt, R.M. and Pastan, I. (1978), *Nature,* **272**, 68–70.

Pruss, R.M., Herschman, H.R. and Klement, V. (1978), *Nature,* **274**, 272–274.

Rees, A.R., Adamson, E.D. and Graham, C.F. (1979), *Nature,* **281**, 309–311.

Sahyoun, N., Hock, R.A. and Hollenberg, M.D. (1978), *Proc. Natl. Acad. Sci. USA,* **75**, 1675–1679.

Savage, C.R., Jr., Hash, J.H. and Cohen, S. (1973), *J. Biol. Chem.,* **248**, 7669–7672.

Schechter, Y., Hernaez, L., Schlessinger, J. and Cuatrecasas, P. (1979), *Nature,* **278**, 835–838.

Schlessinger, J., Shechter, Y., Cuatrecasas, P., Willingham, M.C. and Pastan, I. (1978a), *Proc. Natl. Acad. Sci. USA,* **75**, 5353–5357.

Schlessinger, J., Shechter, Y., Willingham, M. and Pastan, I. (1978b), *Proc. Natl. Acad. Sci. USA,* **75**, 2659–2663.

Shechter, Y., Hernaez, L. and Cuatrecasas, P. (1978a), *Proc. Natl. Acad. Sci. USA,* **75**, 5788–5791.

Shechter, Y., Schlessinger, J., Jacobs, S., Chang, K.J. and Cuatrecasas, P. (1978b), *Proc. Natl. Acad. Sci. USA,* **75**, 2135–2139.

Sherwin, S.A., Sliski, A.H. and Todaro, G.J. (1979), *Proc. Natl. Acad. Sci. USA,* **76**, 1288–1292.

Shoyab, M., DeLarco, J.E. and Todaro, G.J. (1979), *Nature,* **279**, 387–391.

Smith C.J., Wejksnora, P.J., Warner, J.R., Rubin, C.S. and Rosen, O.M. (1979), *Proc. Natl. Acad. Sci. USA,* **76**, 2725–2729.

Todaro, G.J. and DeLarco, J.E. (1978), *Cancer Res.,* **38**, 4147–4154.

Todaro, G.J., DeLarco, J.E. and Cohen, S. (1976), *Nature,* **264**, 26–31.

Todaro, G.J., DeLarco, J.E., Marquart, H., Bryant, M.L., Sherwin, S.A. and Sliski, A.H. (1979), *Cold Spring Harbor Conf. Cell Prolif.,* **6**, 113–127.

Todaro, G.J., DeLarco, J.E., Nissley, S.P. and Rechler, M.M. (1977), *Nature,* **267**, 526–528.

Ushiro, H. and Cohen, S. (1980), *J. Biol. Chem.,* **255**, 8363–8365.

Vlodavsky, I., Brown, K.D. and Gospodarowicz, D. (1978), *J. Biol. Chem.,* **253**, 3744–3750.

Willingham, M. and Pastan, I. (1978), *Cell,* **13**, 501–507.

Wrann, M.N. and Fox, C.F. (1979), *J. Biol. Chem.,* **254**, 8083–8086.

Wrann, M., Linsley, P.S. and Fox, C.F. (1979), *FEBS Lett.,* **104**, 415–419.

4 Regulation of Prolactin Receptors in Target Cells

ROBERT P.C. SHIU and HENRY G. FRIESEN

Acknowledgement

R.P.C.S. is a scholar, Medical Research Council of Canada.

Receptor Regulation
(*Receptors and Recognition*, Series B, Volume 13)
Edited by R. J. Lefkowitz
Published in 1981 by Chapman and Hall, 11 New Fetter Lane, London EC4P 4EE

4.1 INTRODUCTION

The diverse actions of prolactin have been reviewed by Bern and Nicoll (1968) and Horrobin (1978). Besides its well-known mammogenic role, prolactin is important in regulating the functions of many organs in many species. In the mammals, the mammary gland is undoubtedly one of the principal target organs of prolactin. In this tissue, prolactin is a lactogenic and a mitogenic hormone (Cowie and Tindal, 1971; Ceriani, 1974). The biological actions of prolactin in other organs such as liver, kidney and the reproductive organs are less well defined. Nevertheless, prolactin has been reported to stimulate somatomedin production (Francis and Hill, 1970) and ornithine decarboxylase activity in rat liver (Richards, 1975) and to affect kidney's ability to retain Na^+ ions (Nicoll, 1975). The luteotropic and luteolytic action in the ovary is well documented (Nicoll, 1975; Wuttke and Meites, 1971). Prolactin also stimulates prostatic growth (Moger and Geschwind, 1972) and uptake of testosterone by the prostate (Farnsworth, 1977).

It is generally accepted that prolactin initiates its action by binding to a specific cell surface receptor. In spite of this general belief, the obligatory role of the prolactin receptor in mediating the biological action of prolactin has only been demonstrated experimentally in the mammary gland (Shiu and Friesen, 1976a). Whether the prolactin binding sites are synonymous with prolactin receptors in other tissues remains to be proven. Nevertheless, the similarity between the prolactin receptors in the mammary tissue and the prolactin binding sites in other tissues is highly suggestive of a 'receptor' function of a 'binding site'. The prolactin binding characteristics of the binding sites in the other organs are identical to that of the mammary prolactin receptors. Using antibodies to the rabbit mammary prolactin receptors, we (Shiu and Friesen, 1976b) found that the prolactin binding sites in many tissues are immunologically very similar to the prolactin receptors in the mammary gland. Moreover, *in vivo* administration of anti-receptor antibodies partially inhibited the lactogenic response of prolactin in lactating rats and effectively blocks the luteolytic effect of prolactin in normal cycling rats (Bohnet *et al.*, 1978). These studies suggest that the prolactin binding sites in most, if not all, tissues are biologically important prolactin receptor sites.

Traditionally, it is thought that the magnitude of biological response in a given tissue is governed only by the concentration of a hormone such as prolactin. Since the discovery of prolactin receptors and the demonstration that the prolactin receptors are dynamically controlled by trophic factors, many believe that prolactin receptors may play a major role in determining the sensitivity of target tissues to prolactin. Therefore, to define better the significance of prolactin in the overall physiology of a living system, we have to understand not only the regulation of production and secretion of prolactin by the pituitary gland, but also the regulation of the prolactin receptors in target tissues. Since Waters *et al.* (1978) have reviewed this area, we will focus on more recent developments on the regulation of prolactin receptors in a number of prolactin target organs.

In the discussion that follows, the term 'prolactin receptors' will be used synonymously with 'lactogenic receptors' because prolactin receptors cannot discriminate between prolactin and other lactogenic hormones that include placental lactogens and human growth hormone, but not non-primate growth hormones, which are not lactogenic. A receptor that binds only growth hormones, but not prolactin, is referred to as a growth hormone receptor; human growth hormone binds to both prolactin receptors and growth hormone receptors. Many tissues possess both types of receptors.

4.2 REGULATION OF PROLACTIN RECEPTORS IN THE MAMMARY GLAND

4.2.1 Normal mammary gland

The mammogenic role of prolactin is its oldest established role and the initial studies on prolactin receptors were carried out in the mammary gland (Turkington *et al.*, 1973; Shiu *et al.*, 1973; Shiu and Friesen, 1974a,b; 1976a,b,c; 1980). Yet changes in prolactin binding in this tissue under various physiological conditions and the influence of trophic factors have not been as fully characterized as in the liver. Nevertheless, Frantz *et al.* (1974) first showed that prolactin binding activity in lactating rat mammary gland is many times higher than that in tissues from pregnant animals. Using rat mammary slices for their studies, Holcomb *et al.* (1977) also observed that prolactin binding remains low until parturition, when a very large (20-fold) increase occurs; this high level of prolactin binding is maintained for two weeks after which a gradual decline occurs. Bohnet *et al.* (1977) observed a similar pattern of prolactin binding in membrane particles prepared from rat mammary glands throughout pregnancy and lactation.

Using collagenase-dispersed cells from mouse mammary gland, Sakai *et al.* (1978) not only confirmed changes in prolactin receptor activity throughout pregnancy and lactation but also showed that epithelial cells from adult virgin animals actually contain the highest prolactin binding activity as compared with cells derived from pregnant and lactating mice. Higher receptor activity in virgin rabbit mammary cells was observed by Suard *et al.* (1979).

A number of reasons can account for the fluctuation in prolactin binding in the mammary gland throughout pregnancy and lactation. It is known that in many species circulating levels of prolactin and placental lactogen are high throughout pregnancy (Kelly *et al.*, 1976a). A large number of prolactin receptors could have been saturated by endogenous hormones and are there-fore not available for binding of radiolabeled prolactin. Indeed, hysterectomy of pregnant rats (to remove placental lactogen) resulted in a large increase in prolactin binding from day 11 onwards (Holcomb *et al.*, 1977) corresponding to the secretion of rat placental lactogen (Shiu *et al.*, 1973). In species, such as the rabbit, which do not secrete placental lactogen, suppression of prolactin secretion *in vivo* by ergocornine 36 hours prior to measurement of prolactin receptors resulted in higher estimates of prolactin binding in mammary glands of the rabbits throughout pregnancy and lactation (Djiane *et al.*, 1977), although prolactin binding determined in this manner in pregnant mammary gland is still lower than that in lactating gland. Receptor occupancy, there-fore, cannot fully account for the lower prolactin binding in the mammary gland from pregnant animals. The circulating level of progesterone is high during the later stages of pregnancy and prolactin surges at parturition. This led some investigators to speculate that progesterone is responsible for sup-pressing prolactin receptors and that prolactin is responsible for inducing its own receptors in the mammary gland, as it seems to do for the hepatic prolactin receptor (to be discussed in later sections). Hence, Bohnet *et al.* (1977) were able to block the post partum increase in prolactin receptors in the rat mammary gland, if post partum prolactin surge is blocked by ergocor-nine prior to parturition or by the removal of suckling stimulus (removal of pups) at delivery. Djiane and Durand (1977) also observed that injection of prolactin induces prolactin receptors in pseudopregnant rabbit mammary gland and this induction was abolished by progesterone injection. This would extend the well-known inhibitory effect of progesterone on lactogenesis to the level of the prolactin receptor. The mechanism by which progesterone sup-presses prolactin receptor in the mammary gland is not known. However, Sakai *et al.* (1978) suggested that progesterone competes with the cytoplasmic glucocorticoid receptors (Shyamala, 1975) resulting in the suppression of glu-cocorticoid-induced replenishment of prolactin receptors (Sakai *et al.*, 1979).

The nature of prolactin induction of its own receptors is not clear. [Recep-tors for prolactin and angiotensin II are the only peptide hormone receptors that are 'up-regulated' after exposure to elevated hormone concentrations; most other peptide hormone receptors are down-regulated by elevated hor-mone concentrations (Catt *et al.*, 1979).] In view of the recent observations that receptor-bound prolactin is internalized by mammary epithelial cells (Shiu, 1980; Suard *et al.*, 1979) it has been suggested that the internalized prolactin or receptor or fragments derived from these molecules may signal the turning on of the synthesis of receptors. This is just a speculation and no experimental data are yet available to support this notion.

Thyroid hormones are known to potentiate the effect of prolactin on stimulating milk protein synthesis in the mammary gland, possibly by sensitizing the mammary cells to prolactin (Vonderhaar, 1977). The studies by Bhattacharya and Vonderhaar (1979) showed that mammary glands obtained from hypothyroid mice contain reduced content of prolactin receptors. Using mammary organ cultures, they were able to show that T_3 induces prolactin receptors by about 2-fold and simultaneously renders T_3-treated tissues more sensitive to prolactin in inducing the activity of lactose synthetase activity. More interestingly, they showed that puromycin, an inhibitor of protein synthesis, does not abolish the inductive effect of prolactin receptors by T_3 although the inhibitor abolishes synthesis of cellular proteins by 95 per cent. This, it appears that in short-term cultures, T_3 acts by activating or unmasking pre-existing receptor molecules. The modulation of prolactin binding by surface-active agents such as concanavalin A (Con A) was also reported by Costlow and Gallagher (1977) who demonstrated that Con A inhibits binding of prolactin to a number of tissues, including the rat mammary gland. α-Methylmannoside and α-methylglucoside, which inhibit the binding of Con A, reversed the Con A inhibition of prolactin binding. Conceivably, Con A cross-links membrane glycoproteins [prolactin receptor is a glycoprotein (Shiu and Friesen, 1980)] and aggregated prolactin receptors probably fail to recognize prolactin or have lower affinity for prolactin. It is also possibly that prolactin receptors cross-linked by Con A are internalized by the cells, thus effectively reducing the content of prolactin receptors.

In the normal, differentiated mammary gland, the changes in prolactin receptor activity seem to correlate with the biological response of the tissue to prolactin. However, it is important to evaluate quantitatively the magnitude of response for a wide range of receptor contents. Furthermore, the molecular mechanisms by which prolactin receptors are modulated by various trophic factors have to be delineated in order to fully understand the mechanism of action of prolactin in its principal target organ, the mammary gland.

4.2.2 Neoplastic mammary gland

There is good evidence that prolactin promotes growth of mammary tumors in rodents (Welsch and Nagasawa, 1977). Prolactin receptors are readily demonstrable in carcinogen-induced rat mammary tumors. Accordingly, this model has been used to study the role of prolactin and other hormones such as insulin and estrogen in the etiology of breast cancer. A number of studies revealed that rat tumors which grow in response to prolactin administration possess higher levels of prolactin receptor than prolactin-unresponsive tumors, if prolactin receptor activity is assessed after prolactin treatment (Kelly *et al.*, 1974a; Turkington, 1974). These findings could be explained by the fact that prolactin induces its own receptors. Holdaway and Friesen (1976) expanded these studies in 7,12-dimethylbenzanthracene (DMBA)-induced rat mammary tumors by showing no significant difference in prolactin binding in prolactin responding and non-responding (autonomous) tumors prior to treatment, even though such a difference was seen following treatment. However, those tumors with higher prolactin receptor content were likely to regress following suppression of prolactin by bromoergocryptine. Estrogen and androgen also affect growth of rat mammary tumors. Very high doses of estrogen could inhibit tumor growth and prolactin binding (Kledzik *et al.*, 1976a) despite prolactin concentration being elevated. Testosterone injection causes regression in tumors and a 60 per cent reduction in prolactin binding (Costlow *et al.*, 1976). The precise relationship between prolactin receptor and prolactin-dependent growth in rat mammary tumors remains to be established.

Studies on prolactin binding in human breast cancer cells are still in their very infant stages. Limited studies using breast cancer biopsy specimens revealed that a small percentage of tumors contain specific prolactin receptors (Holdaway and Friesen, 1977). However, their clinical significance is not known. No study on regulation of prolactin receptor in human cells has yet been performed. Studies on the significance of prolactin binding to human breast cancer cells and their regulation by trophic factors depend on the availability of some suitable human cell models. Perhaps the use of human breast cancer cells maintained in tissue culture will facilitate these studies. Indeed, we have shown that a number of human breast cancer cell lines maintained in long term tissue culture retain specific binding of human prolactin (Shiu, 1979). These cell lines may potentially be useful for elucidating the biological significance of prolactin receptors and their regulation in human breast cancer cells.

4.3 REGULATION OF HEPATIC PROLACTIN (LACTOGENIC) RECEPTORS

The regulation of prolactin binding in the liver has been studied in greater detail than in any other tissue, primarily due to ease of obtaining large amounts of this tissue. Looking back, perhaps it was a bad choice in using the liver to study the regulation of prolactin binding because the biological significance of prolactin in the liver has not been well defined, as it has for the mammary gland. Perhaps we should have spent more effort in elucidating the actions of prolactin in the liver before spending so much effort in studying prolactin binding. Further, there is no experimental evidence to suggest that the prolactin binding sites in the liver are the 'true' receptor sites which, by definition, should be responsible for initiating the biological effect of prolactin. Nevertheless, since the prolactin binding sites in the liver are immunologically related to the mammary gland prolactin receptors and are dynamically regulated by a wide variety of trophic factors, the chances are that these binding sites are biologically active receptors.

The first evidence to indicate that prolactin receptors are dynamic molecules came from studies using livers from rats in different stages of development. We first observed that a nine-fold increase in prolactin binding capacity occurs between 20 and 40 days of age in female rat liver, with a further two-fold increase during pregnancy (Kelly *et al.*, 1974b); day 40 corresponds to the onset of puberty in the rat and is preceded by increases in prolactin and estrogen. In the normal cycling rat, prolactin binding is significantly higher during estrus and diestrus I (Kelly *et al.*, 1975), just after prolactin, gonadotropin and estradiol surge on proestrus. Prolactin receptor activity in adult male rat liver remains low, at a level similar to that of prepubertal female rats (Kelly *et al.*, 1974b).

On examining other species for developmental changes in lactogenic receptors, a slightly different pattern emerges. In the mouse there is an increase of lactogenic receptors in the female liver upon puberty. The adult female has a higher level of prolactin binding than the male. However, there is no further induction of prolactin receptors during pregnancy but rather, there is an induction of growth hormone specific sites (Posner, 1976a). In the adult rabbit, the hepatic growth hormone receptors out-number the lactogenic receptors; there is no sex difference in lactogenic receptors, but growth hormone receptor in the female liver is higher than that in the male. Again, pregnancy results in the induction of growth hormone receptors but not the prolactin receptors (Kelly *et al.*, 1974b). In guinea pigs, an entirely different pattern was observed. There is no significant binding of either growth hormone or prolactin (Posner *et al.*, 1974; Kelly *et al.*, 1974b), although insulin binding is present in most tissues and increases during pregnancy in the liver (Kelly *et al.*, 1974b). Hence there is a considerable species variation in the induction of hepatic prolactin and growth hormone receptors.

The changes in prolactin receptors in the liver seem to be related to the hormonal status of the animal: sex differences and pregnancy are directly related to hormonal changes. This prompted us to examine the role of various sex-related protein and steroid hormones in the regulation of prolactin receptors in the rat liver. Injection of estradiol or estrone (but not estriol) in male rats for 8–12 days resulted in a marked increase (>10-fold) in hepatic prolactin receptors (Posner *et al.*, 1974b). In adult female animals, this regimen augmented prolactin binding to a level seen in pregnancy. Estrogen-induction of hepatic prolactin binding can also be observed in prepubertal male and female rats. Administration of the male sex steroid, testosterone, led to a disappearance of receptors in the female. Castration in the male led to a substantial increase in hepatic prolactin receptors, presumably due to the removal of androgens, and the administration of testosterone prevented this increase (Aragona and Friesen, 1975).

It is well known that estrogen administration stimulates prolactin secretion by the pituitary. The role of the pituitary and the possible induction of hepatic prolactin receptors by estrogen via the production of prolactin were examined. Hence hypophysectomy decreases prolactin binding and abolishes the ability of estrogen to induce prolactin receptors (Posner *et al.*, 1974b). These observations imply that prolactin itself is the inducer of hepatic prolactin receptors. This view was supported by the observation that renal pituitary implants (which secrete mainly prolactin) in hypophysectomized female rats were able to reduce the loss of lactogenic receptors seen after hypophysectomy in the female and in fact, to induce prolactin receptors in hypophysectomized males (Posner *et al.*, 1975). Implantation of prolactin-secreting tumors greatly elevated circulating levels of prolactin and hepatic lactogenic receptors in both normal male and female rats and prevented the decline of receptors in hypophysectomized animals (Posner, 1976b). Injection of high doses of prolactin (0.5–2 mg) was able to partially (20–30 per cent) restore prolactin receptors in hypophysectomized rats (Costlow *et al.*, 1975; Bohnet *et al.*, 1976). It should be emphasized that prolactin binding reported for these experiments did not take into account receptor occupancy by the high level of prolactin produced by pituitary implants and prolactin injection. If receptor occupancy did occur, then the above values of prolactin binding would have been under-estimated.

These observations led many investigators to believe that the pituitary factor that is responsible for the induction of lactogenic receptors in the liver is prolactin itself. However, there is compelling evidence to suggest that prolactin itself may not be sufficient. Firstly, to induce receptors (in hypophysectomized female rats) by injection of prolactin *in vivo*, only very large doses of prolactin (2 mg) are effective. A small amount of contaminants in the prolactin preparation used could have accounted for the induction; this possibility has not been ruled out. Secondly, when renal pituitary transplants and prolactin-secreting pituitary tumors were used for receptor induction, it has been *assumed* that they secrete only prolactin. It is quite likely that other hormones are secreted by the transplants in variable amounts; these 'other' hormones might have contributed to the induction process. Thirdly, Aragona *et al.* (1976a) and Bohnet *et al.* (1976) have shown that ACTH and growth hormone synergized with prolactin in the induction of hepatic receptor. Moreover, Gelato *et al.* (1975) demonstrated that thyroidectomy reduced the number of prolactin binding sites in the female rat liver, and thyroxine provided effective replacement therapy. These above findings suggest that other pituitary hormones (i.e. ACTH, growth hormone and TSH) are synergizing factors with prolactin in inducing prolactin receptors. Fourthly, if bromoergocryptine (CB-154) was used to suppress prolactin secretion in ovariectomized rats, the stimulation of hepatic receptors due to estrogen administration was still observed (Kelly *et al.*, 1976b); estrogen is ineffective in hypophysectomized rats. Furthermore, Aragona *et al.* (1976b) showed that castration in male rats resulted in an increase of hepatic prolactin receptors, and this induction is pituitary dependent. This induction, however, did not appear to be directly mediated by prolactin since serum prolactin was decreased following castration. Moreover, suppression of circulating prolactin to low levels with CB-154 or sustained elevation of prolactin with fluphenazine enanthate did not affect castration-induced increase in prolactin binding. In fact, testosterone suppression of receptor induction that follows castration was associated with elevated serum prolactin level! Sherman *et al.* (1977) also showed that testosterone lowered prolactin binding in female rat livers without changing prolactin levels. Estrogen increased both plasma prolactin and prolactin binding, the estrogen effect on plasma prolactin was not blocked by testosterone, but the effect on prolactin binding in the liver was, suggesting that induction of prolactin binding by estrogen and inhibition of binding by testosterone was not mediated by prolactin. Finally, prolactin administered to hypophysectomized or normal male rats was unable to induce liver lactogenic receptors (Costlow *et al.*, 1975; Aragona *et al.*, 1976a), even when serum prolactin levels exceeded those reported for hypophysectomized male rats bearing renal pituitary implants or prolactin-secreting tumors.

All the above observations seem to suggest to us that prolactin *may* not be the inducer of prolactin receptors. We wish to hypothesize that there is an as yet unidentified 'factor' in the pituitary that is responsible for the induction of prolactin receptor in the liver. This 'factor' may be identical to the 'feminizing factor' or 'feminotropin' from the pituitary described by Gustafsson and co-workers (Stenberg *et al.*, 1977; Eneroth *et al.*, 1976). The production of this pituitary factor was increased by estrogen and was suppressed by testosterone. This factor would have been present as a contaminant in the prolactin (and other hormones) preparations used for receptor induction experiments. The fact that only large (mg) doses of prolactin were effective seems to favor our 'contaminant' hypothesis. Finally, this pituitary factor was secreted by pituitary implants and other pituitary tumors.

4.4 REGULATION OF PROLACTIN BINDING IN THE MALE AND FEMALE REPRODUCTIVE ORGANS

Effects of trophic factors in influencing prolactin binding have been examined in a number of reproductive organs in the rat. In the male, the ventral prostate possesses the highest prolactin binding activity and the epididymis, seminal vesicles and testis have lower, but significant, binding of prolactin (Aragona and Friesen, 1975).

Prostatic binding of prolactin parallels that of testosterone secretion; both activities are highest between days 20 and 70 and decrease to 15 percent of this level by day 270 (Aragona *et al.*, 1977; Barkey *et al.*, 1977). Again, prostatic prolactin binding is sensitive to a testosterone–estrogen balance, but in a reverse fashion to the liver. Thus, administration of estrogen or castration decreased prostatic prolactin binding whereas testosterone replacement restored prolactin binding in prostates in castrated animals (Kledzik *et al.*, 1976b; Charreau *et al.*, 1977).

Prolactin synergizes with LH in promoting testosterone production by Leydig cells; suppression of prolactin secretion by CB-154 decreases testosterone level (Hafiez *et al.*, 1972). Prolactin binding has been localized to the interstitial cell fractions, perhaps exclusively to the Leydig cells (Aragona *et al.*, 1977; Charreau *et al.*, 1977) in the testis. Aragona *et al.* (1977) found that CB-154 had no effect on prolactin binding but, surprisingly, lowered LH binding in the Leydig cells. This may explain the synergistic effect of prolactin on LH-induced steroidogenesis in Leydig cells.

In female rats, the luteotropic and luteolytic actions of prolactin have been well documented (Nicoll, 1975; Wuttke and Meites, 1971). Accordingly, prolactin binding in the ovary is readily demonstrable (Shiu and Friesen, 1974a; Posner *et al.*, 1974a). Using FSH–estradiol primed hypophysectomized immatured rats, Richards and Williams (1976) found that LH injection resulted in an increase in prolactin receptor activity in luteal cells. Prolactin was able to stimulate progesterone production 48 hours after LH injection, provided prolactin treatment was initiated at the same time as LH. This is an example of induction of prolactin receptors by another hormone being followed closely by a biological response to prolactin. If, on the other hand, prolactin was administered 72 hours after LH injection, then prolactin injection led to luteolysis, rather than luteotrophy. If the prolactin receptors mediate both the luteolytic and luteotrophic effect of prolactin, then other trophic factors must be the deciding factors in determining whether prolactin is luteotrophic or luteolytic. Interestingly, as a result of administration of antiserum against prolactin receptors into normal cycling rat, we found (Bohnet *et al.*, 1978) that only the luteolytic effect of prolactin was blocked by the antibodies to the receptors.

4.5 REGULATION OF PROLACTIN BINDING IN THE KIDNEY AND ADRENAL GLAND

One of the functions of prolactin is the regulation of electrolyte balance in vertebrates (Bern and Nicoll, 1968; Nicoll, 1975). This may be accomplished by the direct action of prolactin on the target tissue, as seems to be the case in mammary tissue where prolactin stimulates Na^+ retention (Falconer and Rowe, 1977; Bisbee *et al.*, 1979). Prolactin may also influence electrolyte balance via its ability to induce mineralocorticoid production by the adrenal (Witorsch and Kitay, 1972). Prolactin binding is high in the adrenal and kidney (Shiu and Friesen, 1974a; Posner *et al.*, 1974a), and salt-loading significantly increased prolactin binding in the adrenal but not in the kidney (Marshall *et al.*, 1975). Prolactin binding in the adrenal may also reflect an action of prolactin on metabolism of steroids other than the mineralocorticoids. Prolactin binding in adrenal and kidney is modulated by estrogen and testosterone in a manner similar to that found in the prostate gland: testosterone increases while estrogen decreases prolactin binding (Monkemeyer and Friesen, 1974). However, the physiological significance of regulation of prolactin binding by these hormones is not known.

4.6 CONCLUSION

Prolactin receptors in many target organs are dynamically regulated by a wide variety of trophic factors. The ability of steroid hormones to modulate the action of prolactin is mediated in part by changes in the prolactin receptors. A number of pituitary hormones such as adrenocorticotropin, growth hormone, thyroid stimulating hormone, gonadotropins and of course prolactin itself are implicated in the regulation of prolactin receptors. A 'feminizing' factor from the pituitary gland may play a major role in prolactin receptor regulation, although the identity of this factor remains to be established. The interaction of all the known trophic factors which have been shown to regulate prolactin receptors in a number of prolactin target tissues is summarized in Fig. 4.1.

One of the major obstacles in these studies is the lack of suitable model systems which would allow the titration of receptor occupancy and biological response to prolactin. The mammary gland and, to a lesser extent, the corpus luteum are the only situations where prolactin receptor concentration is correlated with biological response to prolactin. In most of the receptor induction experiments, it is not known whether the trophic factor in question exerted direct or indirect effects. We feel that effort should be spent on finding some *in vitro* model systems that will enable us to examine the direct effects of some of these trophic factors. Further, the *in vitro* models would possibly enable us to investigate the mechanisms of prolactin receptor induction and reduction at the molecular level. One requirement of any *in vitro* cell model is that these cells have to be hormonally responsive. To obtain viable prolactin target cells that are hormonally responsive *in vitro* has been the major objective of many laboratories.

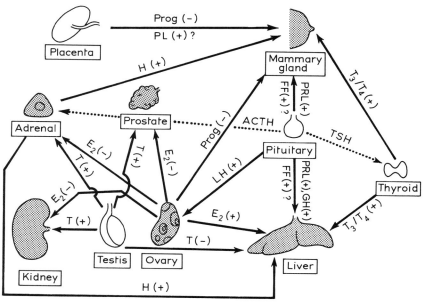

Fig. 4.1 Regulation of prolactin receptors in some target organs by positive and negative trophic factors: shaded organs are those in which prolactin receptors were measured. $(+)$, positive and $(-)$, negative trophic effects. ACTH, adrenocorticotropic hormone; E_2, estradiol; FF, 'Feminizing factor' or 'feminotropin'; GH, growth hormone, H, hydrocortisone; LH, luteinizing hormone; PRL, prolactin; PL, placental lactogen; PROG, progesterone; T, testosterone; T_3/T_4, triiodothyronine/thyroxine; TSH, thyroid stimulating hormone.

REFERENCES

Aragona, C., Bohnet, H.G., Fang, V.S. and Friesen, H.G. (1976a), *Endocrine Res. Comm.*, **3**, 199–208.

Aragona, C., Bohnet, H.G. and Friesen, H.G. (1976b), *Endocrinology*, **99**, 1017–1022.

Aragona, C., Bohnet, H.G. and Friesen, H.G. (1977), *Acta Endocrinol.*, **84**, 402–409.

Aragona, C. and Friesen, H.G. (1975), *Endocrinology,* **97**, 677–684.

Barkey, R.J., Shani, J., Amit, T. and Barzilai, D. (1977), *J. Endocrinol.*, **74**, 163–173.

Bern, H.A. and Nicoll, C.S. (1968), *Rec. Prog. Horm. Res.*, **24**, 681–720.

Bhattacharya, A. and Vonderhaar, B.K. (1979), *Biochem. Biophys. Res. Commun.*, **88**, 1405–1411.

Bisbee, C.A., Machen, T.E. and Bern, H.A. (1979), *Proc. Natl. Acad. Sci. USA*, **76**, 536–540.

Bohnet, H.G., Aragona, C. and Friesen, H.G. (1976), *Endocrine Res. Comm.,* **3,** 187–198.

Bohnet, H.G., Gomez, F. and Friesen, H.G. (1977), *Endocrinology,* **101,** 1111–1121.

Bohnet, H.G., Shiu, R.P.C., Grinwich, D. and Friesen, H.G. (1978), *Endocrinology,* **120,** 1657–1661.

Catt, K.J., Harwood, J.P., Aguilera, G. and Dufau, M.L. (1979), *Nature,* **280,** 109–116.

Ceriani, R.L. (1974), *J. Invest. Dermat.,* **63,** 93–108.

Charreau, E.H., Attramadal, A., Torjesen, P.A., Calandra, R., Purvis, K. and Hansson, V. (1977), *Mol. Cell Endorcinol.,* **7,** 1–7.

Costlow, M.E., Buschow, R.A. and McGuire, W.L. (1975), *Life Sci.,* **17,** 1457–1461.

Costlow, M.E., Buschow, R.A. and McGuire, W.L. (1976), *Cancer Res.,* **36,** 3324–3329.

Costlow, M.E. and Gallagher, P.E. (1977), *Biochem. Biophys. Res. Commun.,* **77,** 905–911.

Cowie, A.Y. and Tindal, J.S. (1971), *Physiology of Lactation*, Edward Arnold (Publishers) Ltd., London.

Djiane, J. and Durand, P. (1977), *Nature,* **266,** 614–616.

Djiane, J., Durand, P. and Kelly, P.A. (1977), *Endocrinology,* **100,** 1348–1356.

Eneroth, P., Gustafsson, J.A., Larsson, A., Skett, P., Stenberg, A. and Sonnerschein, C. (1976), *Cell,* **7,** 413–417.

Falconer, I.R. and Rowe, J.M. (1977), *Endocrinology,* **101,** 181–186.

Farnsworth, W.E. (1972), *Prolactin Carcinogenesis, Proc. 4th Tenovus Workshop* (Boyns, A.R. and Griffiths, K., eds.), Alpha Omega, Cardiff, pp. 217–225.

Francis, M.J.O. and Hill, D.J. (1970), *Nature,* **255,** 167–168.

Frantz, W.L., MacIndoe, J.H. and Turkington, R.W. (1974), *J. Endocrinol.,* **60,** 485–497.

Gelato, M., Marshall, S., Boudreau, M., Bruni, J., Campbell, G.A. and Meites, J. (1975), *Endocrinology,* **96,** 1292–1296.

Hafiez, A.A., Lloyd, C.W. and Bartke, A. (1972), *J. Endocrinol.,* **52,** 327–332.

Holcomb, H.H., Costlow, M.E., Buschow, R.A. and McGuire, W.L. (1977), *Biochim. Biophys. Acta,* **428,** 104–112.

Holdaway, I.M. and Friesen, H.G. (1976), *Cancer Res.,* **36,** 1562–1567.

Holdaway, I.M. and Friesen, H.G. (1977), *Cancer Res.,* **37,** 1946–1952.

Horrobin, D.F. (1978), *Prolactin,* **6,** Eden Press, Montreal.

Kelly, P.A., Bradley, C., Shiu, R.P.C., Meites, J. and Friesen, H.G. (1974a), *Proc. Soc. Exp. Biol. Med.,* **146,** 816–819.

Kelly, P.A., Ferland, L., Labrie, F. and Delean, A. (1976b), *Clin. Res.,* **23,** 614A.

Kelly, P.A., Posner, B.I. and Friesen, H.G. (1975), *Endocrinology.,* **97,** 1408–1415.

Kelly, P.A., Posner, B.I., Tsushima, T. and Friesen, H.G. (1974b), *Endocrinology,* **96,** 532–539.

Kelly, P.A., Tsuchima, T., Shiu, R.P.C. and Friesen, H.G. (1976a), *Endocrinology,* **99,** 765–774.

Kledzik, G.S., Bradley, C.J., Marshall, S., Campbell, G.A. and Meites, J. (1976a), *Cancer Res.,* **36,** 3265–3268.

Kledzik, G.S., Marshall, S., Campbell, G.A., Gelato, M. and Meites, J. (1976b), *Endocrinology,* **98,** 373–379.

Marshall, S., Gelato, M. and Meites, J. (1975), *Proc. Soc. Exp. Biol. Med.*, **149**, 185–188.

Moger, W.H. and Geschwind, I.I. (1972), *Proc. Soc. Exp. Biol. Med.*, **141**, 1017–1021.

Monkemeyer, H. and Friesen, H.G. (1974), *Clin. Res.*, **22**, 733A.

Nicoll, C.S. (1975), *Handbook Physiol., Section 7: Endocrinology*, **4**, 253–292.

Posner, B.I. (1976a), *Endocrinology*, **98**, 645–654.

Posner, B.I. (1976b), *Endocrinology*, **99**, 1168–1177.

Posner, B.I., Kelly, P.A. and Friesen, H.G. (1974b), *Proc. Natl. Acad. Sci. USA*, **71**, 2407–2410.

Posner, B.I., Kelly, P.A. and Friesen, H.G. (1975), *Science*, **188**, 57–59.

Richards, J.F. (1975), *Biochem. Biophys. Res. Commun.*, **63**, 292–299.

Richards, J.S. and Williams, J.J. (1976), *Endocrinology*, **99**, 1571–1581.

Sakai, S., Bowman, P.D., Yang, J., McCormick, K. and Nandi, S. (1979), *Endocrinology*, **104**, 1447–1449.

Sakai, S., Enami, J., Nandi, S. and Banerjee, M.R. (1978), *Mol. Cell Endocrinol.*, **12**, 285–298.

Sherman, B.M., Stagner, J.I. and Zamudio, R. (1977), *Endocrinology*, **100**, 101–107.

Shiu, R.P.C. (1979), *Cancer Res.*, **39**, 4381–4386.

Shiu, R.P.C. (1980), *J. Biol. Chem.*, **255**, 4278–4281.

Shiu, R.P.C. and Friesen, H.G. (1974a), *Biochem. J.*, **140**, 301–311.

Shiu, R.P.C. and Friesen, H.G. (1974b), *J. Biol. Chem.*, **249**, 7902–7911.

Shiu, R.P.C. and Friesen, H.G. (1976a), *Science*, **192**, 259–261.

Shiu, RP.C. and Friesen, H.G. (1976b), *Biochem. J.*, **157**, 619–626.

Shiu, R.P.C. and Friesen, H.G. (1976c), *Methods Mol. Biol.*, **9**, 565–598.

Shiu, R.P.C. and Friesen, H.G. (1980), *Ann. Rev. Physiol.*, **42**, 83–96.

Shiu, R.P.C., Kelly, P.A. and Friesen, H.G. (1973), *Science*, **180**, 968–971.

Shyamala, G. (1975), *Biochemistry*, **14**, 437–444.

Stenberg, A., Eneroth, P., Gustafsson, J.-A. and Skett, P. (1977), *J. Steroid Biochem.*, **8**, 603–607.

Suard, Y.M.L., Kraehenbuhl, J.-P. and Aubert, M.L. (1979), *J. Biol. Chem.*, **254**, 10466–10475.

Turkington, R.W., Majumder, G.C., Kadohama, N., MacIndoe, J.H. and Frantz, W.L. (1973), *Rec. Prog. Horm. Res.*, **29**, 417–449.

Turkington, R.W. (1974), *Cancer Res.*, **34**, 758–763.

Vonderhaar, B.K. (1977), *Endocrinology*, **100**, 1423–1431.

Waters, M.J., Friesen, H.G. and Bohnet, H.G. (1978), *Receptors and Hormone Action III* (Birnbauer, L. and O'Malley, B.W., eds), Academic Press, New York, pp. 457–477.

Welsch, C.W. and Nagasawa, H. (1977), *Cancer Res.*, **37**, 951–963.

Witorsch, R.J. and Kitay, J.I. (1972), *Endocrinology*, **91**, 764–769.

Wuttke, W. and Meites, J.(1971), *Proc. Soc. Exp. Biol. Med.*, **137**, 988–991.

5 Regulation of Adrenergic Receptors

ALBERT O. DAVIES and ROBERT J. LEFKOWITZ

Receptor Regulation
(*Receptors and Recognition*, Series B, Volume 13)
Edited by R. J. Lefkowitz
Published in 1981 by Chapman and Hall, 11 New Fetter Lane, London EC4P 4EE
© 1981 Chapman and Hall

5.1 INTRODUCTION

Transmission of intercellular messages by the adrenergic mediators, epinephrine and norepinephrine, is a ubiquitous phenomenon in vertebrates which continues to invoke investigative effort after many years of research. It has become apparent that both the release and biological actions of these mediators are modulated very specifically by pharmacological and pathological states. An understanding of the mechanisms of modulation of adrenergic actions is fundamental to addressing the relationships between adrenergic responses and other hormonal or cellular responses.

The development of the basic concepts of adrenergic action such as the notions of agonists, antagonists, intrinsic activity, rank order potency series, etc. bring us to the brink of understanding hormone action. However, a complete understanding has been impeded by the paucity of details known about the molecular basis of these concepts. Early pharmacological and physiological data suggested the existence of discrete cellular sites, termed receptors, which transduce the adrenergic message into cellular actions. Radioligand binding techniques have identified and characterized these receptors and now allow more refined biochemical examinations of the molecular mechanisms of adrenergic hormone action.

Extensive data has accumulated on the adrenergic receptors ranging from their biochemical characteristics to their clinical implications. In this chapter we will focus primarily upon the process of regulation of the adrenergic receptors with an emphasis on regulatory influences that may be of broad interest to investigators in the fields of biochemistry, pharmacology, physiology and medicine. This chapter represents an update of a recent review (Davies and Lefkowitz, 1980a) of this very rapidly moving field. Other selected reviews are referenced which allow the interested reader to pursue related topics in greater depth.

5.2 BACKGROUND

5.2.1 Physiological foundations

Fundamental observations on the specificity of adrenergic agonists were reported in 1948 by Ahlquist, who examined the physiological response of several tissues to catecholamines. Prior to that time, it had been thought that sympathetic actions could be produced by two substances, Sympathin E and Sympathin I, which mediated excitation or inhibition, respectively. Later based largely on Ahlquist's work, this concept was modified to state that adrenergic action could be explained by two classes of receptors, one that was excitatory and another that was inhibitory. Ahlquist demonstrated that the earlier concepts were incorrect by determining a single rank order potency series for a group of adrenergic agonists (l-epinephrine > d,l-epinephrine > norepinephrine > methylnorepinephrine > methylepinephrine > N-isopropyl-epinephrine) applied to a set of adrenergic effects: vasoconstriction, uterine and ureteral contraction, contraction of the nictitating membrane, pupillary dilation and gut inhibition. A distinctly different rank order potency series (N-isopropylepinephrine > l-epinephrine > methylepinephrine > d,l-epinephrine > methylnorepinephrine > norepinephrine) applied to a second set of adrenergic effects: vasodilation, uterine inhibition, and myocardial stimulation. Hence, he postulated that these tissue-specific effects could be explained by two types of receptors, which he termed alpha and beta, respectively, either of which could have inhibitory or stimulatory effects depending on the tissue in question.

Later, by applying a similar approach among beta-receptors, Lands *et al*. (1967) showed that within the beta rank order potency series, two distinct sub-patterns could be distinguished. Myocardial stimulation and lipolysis were associated with a potency series of epinephrine $\geq$ norepinephrine. Bronchial relaxation, vasodilation, and uterine inhibition were associated with a potency series of epinephrine $\gg$ norepinephrine. The former actions were said to be mediated by a beta$_1$-receptor subtype, and the latter by a beta$_2$-receptor subtype. Beta-antagonists may also demonstrate a degree of subtype specificity, with metoprolol preferentially blocking beta$_1$-receptor and butoxamine blocking beta$_2$-receptors. However, this preference is neither absolute nor completely established.

The alpha-receptor system also has been shown to consist of at least two receptor subtypes. For a detailed discussion of the physiological basis of this distinction, the reader is referred to the works of Langer (1974) and Starke *et al*. (1975). Briefly stated, it was demonstrated that a tissue, rabbit pulmonary artery for example, could contain an alpha-receptor which mediates effects on smooth muscle contraction as well as a distinct noradrenergic receptor site that mediates inhibition of neurotransmitter release by the sympathetic nerve terminal. The former sites have generally been referred to as postsynaptic receptors while the latter have been called presynaptic receptors. However, to date, no anatomical data rigorously supporting a presynaptic location for these receptors has been developed. Moreover, Berthelson and Pettinger (1977) have demonstrated 'presynaptic' type receptors to be responsible for inhibition of renin release from the kidney, for central alpha-adrenergically-mediated blood pressure depression, and inhibition of melanocyte stimulating hormone-induced granule dispersion. Finally it has been demonstrated that there are alpha-receptors in non-neuronal tissues such as platelets (Hoffman *et al.*, 1979) with many of the same characteristics as the 'presynaptic' receptors. Thus, there is ample justification to abandon the 'pre'- and 'post'-synaptic designations in favor of the terms alpha$_2$- and alpha$_1$-receptors respectively. The former are characterized by more potent inhibition by yohimbine than prazosin, and are similar to those previously termed 'presynaptic'. While the latter are characterized by potent inhibition of norepinephrine action by prazosin rather than yohimbine and are similar to those previously termed 'postsynaptic'. Vasoconstriction appears to be mediated largely via alpha$_1$-receptors, and CNS sympathetic neural inhibition and platelet aggregation appear to be mediated by alpha$_2$-receptors. There are other alpha-adrenergic drugs that may be somewhat selective but the data are not certain. In general, methoxamine may behave as an alpha$_1$-agonist while clonidine may behave as an alpha$_2$-agonist.

Table 5.1 lists the receptor subtypes present in many of the tissues that will be discussed in this review. With radioligand binding techniques it is now possible to quantitate the relative preponderance of receptor subtypes, as seen in the table (see below). In fact, the identification of receptor subtypes may precede physiological identification. For example, the presence of alpha$_2$-receptors in uterus has been shown by direct radioligand binding, but no clear physiological function for these receptors is presently known.

Table 5.1 Receptor subtypes present in representative tissues discussed in this chapter

Tissue	Subtype	Reference
Cardiac Ventricle (rat)	Beta$_1$ (100%)	Hancock *et al.* (1979)
Cardiac Ventricle (frog)	Beta$_1$ (15–25%)	Hancock *et al.* (1979)
	Beta$_2$ (75–85%)	
Adipocyte	Beta$_1$	Lefkowitz *et al.* (1976)
Vascular Smooth Muscle	Beta$_2$	Lefkowitz *et al.* (1976)
Bronchial Smooth Muscle	Beta$_2$	Lefkowitz *et al.* (1976)
Uterine Smooth Muscle	Beta$_2$	Lefkowitz *et al.* (1976)
Erythrocyte (frog)	Beta$_2$	Lefkowitz *et al.* (1976)
Leukocyte (human)	Beta$_2$	Williams *et al.* (1976)
Vascular	Alpha$_1$	Berthelson and Pettinger (1977)
Kidney	Alpha$_2$	Berthelsen and Pettinger (1977)
CNS (↓ BP)	Alpha$_2$	Berthelsen and Pettinger (1977)
Adipocyte	Alpha$_2$	Berthelsen and Pettinger (1977)
Platelet (human)	Alpha$_2$ (100%)	Hoffman *et al.* (1980)
Liver (rat)	Alpha$_1$ (78%)	Hoffman *et al.* (1980)
	Alpha$_2$ (22%)	
Uterus (rabbit)	Alpha$_1$ (35%)	Hoffman *et al.* (1980)
	Alpha$_2$ (65%)	

5.2.2 The second-messenger concept

With the development of the second-messenger concept (Sutherland, 1971) has come the recognition of an association between certain adrenergic responses and alterations in adenylate cyclase activity. It has become apparent that $beta_1$- and $beta_2$-receptor activation is often associated with activation of adenylate cyclase, with resultant rises in cAMP concentrations intracellularly. Further, some $alpha_2$-receptor responses are associated with a fall in adenylate cyclase activity (e.g. in the human platelet) and, therefore, in intracellular cAMP concentration. The ultimate biochemical targets of alpha- and beta-adrenergic receptor activation may include entities such as glycogen phosphorylase (alpha), glycogen synthetase (alpha), and protein kinase (beta).

5.2.3 Radioligand binding studies

Historically, adrenergic receptor identification began with attempts to bind tritiated catecholamine agonists to the beta-receptor, in analogy with efforts to identify other hormone receptors. The early investigations with these radioligands ([^{3}H]catecholamines) were not successful due to their low specific radioactivity and to the propensity of these ligands, under the assay conditions utilized, to bind to non-receptor sites of uncertain significance. It was inferred from pharmacological and physiological studies (see above) that the true adrenergic receptors should demonstrate certain characteristics in binding studies. These include stereospecificity [(−)isomers more potent than (+)isomers], appropriate rank order of potency in competition studies, and saturability (presuming a finite number of receptors), and other properties discussed in detail in the referenced reviews. Early binding experiments (1969–1973) with [^{3}H]catecholamines did not demonstrate all of these characteristics. Attention was then focused on the use of radiolabeled beta-adrenergic antagonists. [^{3}H]Propranolol was among the earliest agents studied (Levitski *et al.*, 1974). However, low specific radioactivity and a very high percentage non-specific binding compromised the utility of this ligand, although results were more promising than those with [^{3}H]catecholamines. 'Specific binding' generally refers to that component of ligand binding which possesses the characteristics of true receptor binding, whereas 'non-specific binding' refers to that component of ligand binding not inhibited by high concentrations of unlabeled adrenergic agents and which does not meet the criteria for receptor binding.

Confident identification of the beta-adrenergic receptors was expedited by the development of high affinity radioligands, such as the beta-antagonist, (−)alprenolol, reduced with tritium to (−)[^{3}H]dihydroalprenolol ([^{3}H]DHA). Lefkowitz *et al.* (1974) demonstrated the appropriate binding of this radioligand to frog erythrocyte membrane receptors. Similarly, Aurbach *et al.* (1974) demonstrated binding of another beta-antagonist, [^{125}I]hydroxybenzylpindolol ([^{125}I]HYP), to turkey erythrocyte beta-receptors. Since then, refinements in techniques using these radioligands with higher specific radioactivity have led to specific identification of beta-receptors in a wide variety of tissues. More recently, a beta-agonist, [^{3}H]hydroxybenzylisoproterenol ([^{3}H]HBI), has been used to investigate agonist-specific binding to the beta-receptor, thus shedding light on the fundamental differences between agonist and antagonist binding to the receptors (Williams and Lefkowitz, 1977b; Wessels *et al.*, 1978).

Similarly, recent efforts have established the characteristics of alpha-receptor binding with the agonists [^{3}H]clonidine, [^{3}H]norepinephrine, and [^{3}H]epinephrine, and the antagonists [^{3}H]dihydroergocryptine ([^{3}H]DHE), [^{3}H]prazosin, and [^{3}H]WB4101 in several tissues.

Quantitative analysis of radioligand binding experiments can be used to derive the number of receptor sites present in a tissue and the affinity of the receptor for the radioligand. The derived values can be used to compare effects upon receptors induced by physiological or pharmacological manipulation. This is done by measuring the specific radioligand binding in the presence of progressively more radioligand. This is termed a saturation curve, since as more radioligand is added, more of the receptor sites are filled. Scatchard analysis of saturation curves results in a transformation of the usually hyperbolic curve to a linear plot of negative slope. The dissociation constant of the receptor for the ligand is the negative inverse of the slope, and the number of receptors is derived from the x-intercept. Recently, Hancock *et al.* (1979) have used a computerized modelling technique based upon the law of mass action which more accurately estimates the receptor number and affinity, and may be used for a more refined investigation into the molecular mechanisms of receptor activation by agonists. This method has been applied to the study of human leukocytic beta-adrenergic receptors, in which the enhanced accuracy of this method is demonstrated (Davies and Lefkowitz, 1980b).

With the radioligands currently available for alpha- and beta-adrenergic receptor identification, the properties of the receptor molecules can be carefully analysed. Solubilization of the beta-adrenergic receptor and progressively improved purification procedures (Caron *et al.*, 1979) by affinity chromotography may soon yield the receptor's chemical identity.

5.2.4 Biochemical endocrinology

Since the beta-adrenergic receptor has been studied more extensively than the alpha, its biochemistry is better understood. The molecular mechanism of catecholamine stimulation of adenylate cyclase involves several necessary components: the receptor (R), a nucleotide regulatory site (N), and the catalytic moiety (C) (Stadel *et al.*, 1980). Curve modelling by computer has demonstrated that a high-affinity complex is formed between agonists and the beta-adrenergic receptor (Kent *et al.*, 1980; Hancock *et al.*, 1979). Indeed, this high-affinity complex appears to be composed of hormone (H), R, and N as a ternary complex, HRN (De Lean *et al.*, 1980). Guanine nucleotides appear to exert their regulatory effects through the N-site by converting the high-affinity state into a low-affinity state of the receptor, resulting in activation of the catalytic moiety of adenylate cyclase and release of bound hormone. Hence, the N-site 'couples' receptor occupation by agonists with activation of C. Formation of the high affinity state is therefore essential for beta-adrenergic actions.

5.3 ADRENERGIC RECEPTOR REGULATION

5.3.1 Description of terms

The phenomenon of denervation hypersensitivity has fascinated physiologists. Pharmacologists have been similarly intrigued by drug tachyphylaxis and have debated over potential mechanisms. With the binding techniques available for the adrenergic receptors, it is now possible to investigate the molecular mechanisms involved in the phenomena of receptor sensitivity modulation. Receptor regulation, as used herein, refers to the processes that modify an agent's ability to bind to a receptor and induce an effect. The regulation of an adrenergic effect may involve changes in receptor affinity or number, altered coupling of receptor to adenylate cyclase or other effector mechanisms, or alterations at more distal steps. Receptor regulation may be demonstrable in a variety of experimental situations, including in intact animals, whole cell suspensions, or in isolated membrane preparations. It may occur upon exposure to the receptor's own specific agonists or antagonists, or be caused by other unrelated agents or events. The former is sometimes termed 'homologous' regulation, and the latter, 'heterologous' regulation. Over the last several years it has been recognized that some forms of regulation that were thought to be heterologous initially actually might be homologous in nature, since a condition such as ageing may cause regulation through release of catecholamines. Hence these classifications are not rigid.

It should be clearly pointed out that down regulation is not synonymous with desensitization. The former refers to a fall in receptor number whereas the latter refers to a fall in the coupled biological response to a hormone stimulus. A receptor system may be down regulated or desensitized, but not necessarily both. Similarly, up regulation and supersensitization are not synonymous. We will now examine these various types of regulatory processes in more detail.

5.3.2 Homologous down regulation of beta-adrenergic receptors and desensitization of adenylate cyclase – the model systems

It has been known for many years that chronic exposure of cells or tissues to beta-adrenergic agonists (but not antagonists) leads to a progressive decline in their ability to stimulate adenylate cyclase, cAMP, and physiological responses. In some cases this effect is highly specific, that is, only the response to catecholamines is attenuated and not that to other hormonal effectors such as PGE_1 or non-specific stimulators of the enzyme such as NaF or nucleotides. In other forms of desensitization a more general defect in hormone responsiveness is observed. As will be discussed further below, it seems likely that differing mechanisms may be operative in these two classes of desensitization.

A very well-studied example of the former phenomenon (loss of only catecholamine responsiveness) is that which occurs in the beta-receptor coupled adenylate cyclase of the frog erythrocyte.

With the development of highly specific radiolabeled beta-adrenergic antagonist ligands [^{125}I]iodohydroxybenzylpindolol (Aurbach *et al.*, 1974) and [^{3}H]dihydroalprenolol (Lefkowitz *et al.*, 1974), it became possible to study receptor changes in desensitization of this adenylate cyclase-coupled system. Mukherjee *et al.* (1975, 1976) demonstrated that chronic administration of beta-adrenergic catecholamines *in vivo* led to a 68 per cent decrease in [^{3}H]dihydroalprenolol binding to frog erythrocyte membranes, as well as a 77 per cent fall in adenylate cyclase response to beta-adrenergic stimulation. Mickey *et al.* (1975, 1976) showed similar decrements in [^{3}H]dihydroalprenolol binding after *in vitro* desensitization of frog erythrocytes with beta-adrenergic agonists. Detailed binding curves showed the diminished [^{3}H]dihydroalprenolol binding to be the result of a decrease in receptor number rather than a change in the affinity of the ligand for binding sites (Mickey *et al.*, 1975, 1976). These studies presented the first direct evidence for an alteration in hormone binding as an explanation for desensitization of adenylate cyclase. A fall in receptor number as assayed by radioligand binding was also found to accompany desensitization of prostaglandin E_1-sensitive adenylate cyclase in frog erythrocytes (Lefkowitz *et al.*, 1977). After incubation of the intact cells with prostaglandin E_1, [^{3}H]prostaglandin E_1 binding, and prostaglandin E_1-stimulated adenylate cyclase in cell lysates showed strikingly similar decrements.

After initial *in vivo* studies documented apparently concordant falls in beta-receptor number and maximum catecholamine-stimulated adenylate cyclase, more extensive investigations were done *in vitro* on the frog erythrocyte system to characterize this phenomenon fully. The alteration in adenylate cyclase stimulation by isoproterenol represents a diminution in maximum catecholamine-stimulated adenylate cyclase activity with no major alteration in the apparent affinity for agonist stimulation of the enzyme. The ability of non-specific stimulators, such as NaF or of other specific hormones such as prostaglandin E_1 to stimulate the enzyme remains unaffected. Only the ability of beta-adrenergic catecholamines to stimulate the enzyme is diminished after exposure to beta-adrenergic agonists.

In an extensive series of experiments reasonable parallels were documented in the characteristics of the process by which the number of beta-adrenergic receptors ($[^3H]$dihydroalprenolol binding sites) was reduced and the process leading to the reduction of catecholamine-stimulated adenylate cyclase (Mickey *et al.*, 1976). Thus, the time courses of the two processes were similar; both processes demonstrated beta-adrenergic specificity with various agonists, leading to desensitization with a potency series comparable to their potencies for the stimulation of adenylate cyclase or occupation of the receptors. Moreover, antagonists such as propranolol and alprenolol, though they occupy the receptors, do not lead to desensitization or the fall in beta-adrenergic receptor number (Mickey *et al.*, 1976). These antagonists, however, block the ability of agonists to lead to desensitization. Thus, they are antagonists of both adenylate cyclase activation and desenstitization.

In addition to the actual fall in receptor number revealed by $[^3H]$dihydroalprenolol binding in desensitized frog erythrocytes, there is at least one other major alteration in the receptors that is associated with the process of desensitization. This seems to be related to a process by which the receptors become uncoupled from adenylate cyclase activation. In order to understand how this is manifested in binding studies it is necessary to digress briefly to describe some new information related to the characteristics of beta-adrenergic agonist binding to receptors in relation to adenylate cyclase activation.

It seems self-evident to state that the fundamental difference between agonist and antagonist drugs is that agonists activate a biological process by virtue of their binding to receptors, whereas antagonists that bind to the same receptors cause no such activation. Thus, there must be fundamental differences in the way in which agonists, as opposed to antagonists, bind to beta-adrenergic receptors. Radioligand-binding studies present some clues to the nature of these differences. Fig. 5.1 presents a typical competition curve of the antagonist $(-)$alprenolol in competition with the radioligand $(-)$ $[^3H]$dihydroalprenolol for binding to the beta-receptors in frog erythrocyte membranes. The data points are actual experimentally determined ones. The smooth curve, however, has been drawn by a recently developed computer modeling system which draws the best theoretical curve through a set of data points based on the law of mass action (Kent *et al.*, 1980). This system also has the capability to determine whether a given set of data points is best fit by the assumption of competitive interaction for a single homogeneous class of binding sites or whether it is necessary to postulate two or three or more classes of binding sites in order to obtain the best fit. In the case shown in Fig.

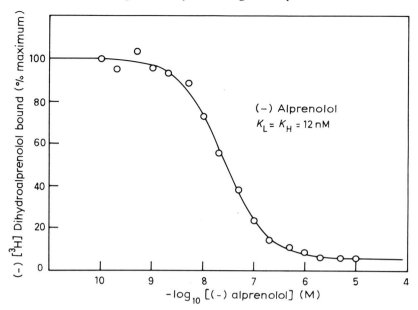

Fig. 5.1 Computerized curve fitting of binding data from displacement of [^{3}H]dihydroalprenolol by (−)-alprenolol (antagonist binding in frog erythrocyte membranes). The abscissa represents the concentration of (−)-alprenolol in the incubation mixture, and the ordinate the bound [^{3}H]DHA. The data points obtained are shown by open circles and represent the mean of duplicate determinations in a single representative experiment. The solid line is a computer-generated curve fitting the observed data points. From Kent *et al.* (1980).

5.1, as well as in all other cases where we have tested antagonists in competition with [^{3}H]dihydroalprenolol, the data points are best fit by a classic single-state binding isotherm. In contrast, the data for agonists are more complex, as shown in Fig. 5.2. Agonist competition curves in this and other systems are often shallow, that is, they have slope factors less than 1. The computer modeling system indicates that the best fit of the experimentally determined data points for the competition of an agonist such as isoproterenol versus [^{3}H]dihydroalprenolol is obtained only upon the assumption of two binding states. One of these has considerably higher affinity than the other. In the example shown in Fig. 5.2 the isoproterenol competition curve appears to be composed of two affinity states, $K_H = 24$ nM (high affinity) and $K_L = 1400$ nM (low affinity). There appear to be approximately 77 per cent of the receptors in the high-affinity state and only 23 per cent in the low-affinity state. Interestingly, in the presence of guanine nucleotides such as GTP and Gpp(NH)p (which are required for hormone activation of adenylate

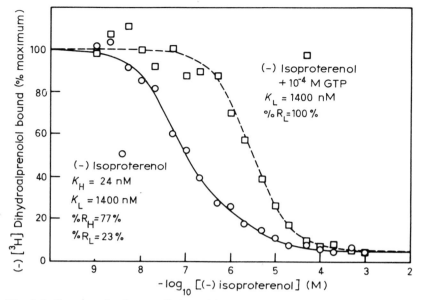

Fig. 5.2 Computerized curve fitting of binding data from displacement of [³H]dihydroalprenolol by (−)-isoproterenol in the presence and absence of guanyl nucleotides in frog erythrocyte membranes. The data points represent the mean of duplicate determinations in a single representative experiment. The solid line is a computer-generated curve fitting the data points in the absence of GTP. The dashed line is a computer-generated curve fitting the data points obtained in the presence of 10^{-4} M-GTP. From Kent *et al.* (1980).

cyclase) the agonist competition curve is shifted to the right and becomes steeper. The curve is now best fit by the assumption of a single state of binding affinity which exactly matches the lower of the two affinity states observed in the absence of guanine nucleotides. These and other data presented elsewhere (Kent *et al.*, 1980) indicate that guanine nucleotides may mediate a transition between the high- and low-affinity state of the receptor. This transition, mediated by guanine nucleotides, appears to occur coincident with the activation of adenylate cyclase by guanine nucleotides. Agonists appear to promote the nucleotide-mediated activation of the enzyme by formation of this high-affinity intermediate binding state. Partial agonists appear to promote formation of this state to a quantitatively smaller extent than do full agonists (Kent *et al.*, 1980). This in turn appears to be related to the fact that the ratio of affinities (K_L/K_H) is smaller for partial than for full agonists. Antagonists do not distinguish between high- and low-affinity states of the receptor, hence do not perturb the equilibrium between such states and do not promote formation of the high-affinity intermediate which appears to be the precursor of adenylate cyclase activation (Kent *et al.*, 1980).

Seen in this light, formation of the high-affinity state of the receptor by agonists is a manifestation of coupling of the receptor to the process of adenylate cyclase activation. Although the exact molecular mechanisms of this coupling process remain to be documented, there are already some interesting clues. Thus, it has been shown that agonists promote an increase in the size of the beta-adrenergic receptor which appears to represent an agonist-induced stabilization of a complex between the receptor and a guanine nucleotide regulatory protein of the adenylate cyclase system (Limbird and Lefkowitz, 1978; Limbird *et al.*, 1980). Thus, this complex of agonist, receptor, and guanine nucleotide regulatory protein, which is of very high affinity and very slowly dissociable, presumably represents an intermediate which dissociates upon interaction with guanine nucleotides.

This type of analysis can be extended to provide additional insights into the abnormalities in the receptor that might be involved in the overall process of desensitization. An example of this approach is provided by the data in Fig. 5.3. Shown here are isoproterenol displacement curves of [^{3}H]dihydroalprenolol binding to frog erythrocyte membranes from control and from previously isoproterenol-desensitized frog erythrocytes. The first and most obvious observation is that the overall amount of [^{3}H]dihydroalprenolol binding is markedly diminished after desensitization. This simply confirms the fact that there are considerably fewer assayable beta-adrenergic receptors present in membranes from the desensitized cells. However, there is another abnormality as well. When computer analysis of the agonist displacement curves is performed, it is found that a smaller percentage of the receptors are present in the high-affinity form, 51 per cent as compared to 81 per cent in this experiment. Fig. 5.4 summarizes average data from five such experiments. There is a statistically significant decrease in the percentage of receptors capable of forming the high-affinity state in the membranes derived from the desensitized as opposed to the control cells. As discussed in detail above, the high-affinity intermediate form of the receptor appears to represent a coupled form of the receptor, formation of which in all likelihood involves the guanine nucleotide regulatory protein. Hence the decrease in high-affinity state receptor in desensitization probably implies that there is some defect in the coupling of the receptors to this guanine nucleotide protein. Thus, there appear to be at least two major defects in beta-adrenergic receptor function in these desensitized cells. The first is a decrease in the overall number of receptors that can be assayed by antagonist ligands such as [^{3}H]dihydroalprenolol. The second is an 'uncoupling' of the receptor, which is apparent as a rightward shift in an agonist competition curve for antagonist binding (Wessels *et al.*, 1978). This effect is resolved by computer modeling procedures into a fall in the ability to form the 'coupled' high-affinity guanine nucleotide-sensitive

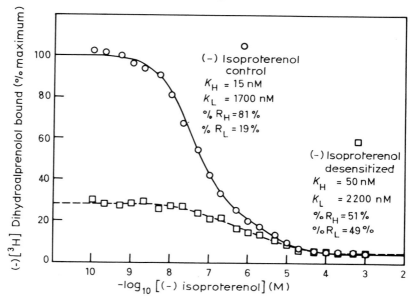

Fig. 5.3 Computerized curve fitting of binding data from displacement of
[³H]dihydroalprenolol by (−)-isoproterenol in control membranes and membranes derived from desensitized cells. Whole frog erythrocytes were desensitized by incubation of the cells with 0.1 mM-(−)-isoproterenol for 3 hours. The
data points obtained for control membranes and for membranes derived from
desensitized cells represent the mean of duplicate determinations in a typical
experiment. The solid and dashed lines are computer-generated curves fitting
the points obtained in these control and desensitized preparations, respectively.
From Kent *et al.* (1980).

intermediate state of the receptor. It is this inability to form the coupled state
of the receptor that is presumably responsible for the previously reported
observation that direct agonist binding as assessed with the agonist
[³H]hydroxybenzylisoproterenol is even more diminished after desensitization of frog erythrocytes than antagonist [³H]dihydroalprenolol binding. This
is because [³H]hydroxybenzylisoproterenol binding appears to measure
largely, if not exclusively, the binding of the agonist to this high-affinity state,
whereas binding of the antagonist [³H]dihydroalprenolol is rather a measure
of the entire pool of beta-adrenergic receptors.

Consistent with the above observations of desensitization in frog erythrocytes is the hypothesis that desensitized receptors arise exclusively from the
high-affinity state of the receptor with chronic occupancy by an agonist.
Indeed, treatment of frog erythrocytes with a chemical intervention such as
N',N'-dicyclohexylcarbodiimide, which prevents formation of the coupled

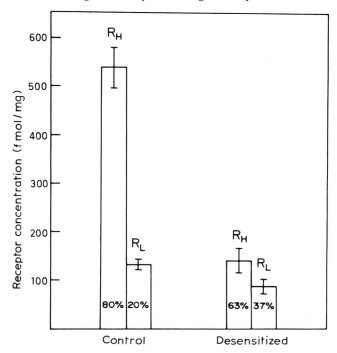

Fig. 5.4 Effect of desensitization on high and low affinity state receptor concentrations. Whole erythrocyte desensitization was carried out as described in the legend for Fig. 5.3. [^{3}H]Dihydroalprenolol-($-$)-isoproterenol competition curves were prepared for control membranes and membranes derived from desensitized cells and the concentrations of high (R_H)- and low (R_L)-affinity state receptor were determined by computer modeling. The results are expressed as the mean ±SEM for five experiments. From Kent *et al.* (1980).

state of the receptor, also prevents the loss of binding sites for [^{3}H]dihydroalprenolol after exposure of the cells to an agonist (Wessels *et al.*, 1979). Although these experiments suggest a role for agonist high-affinity binding in the desensitization process, subsequent data have placed new constraints on the model. Treatment of frog erythrocytes with concentrations of the sulfhydryl reagent *N*-ethylmaleimide, which inactivates adenylate cyclase catalytic activity but which allows high-affinity nucleotide-sensitive agonist binding, also prevents the receptor desensitization process (Stadel and Lefkowitz, 1979). Therefore, agonist occupancy of a nucleotide-sensitive receptor complex is insufficient by itself to initiate down-regulation of receptors in the whole cell. It is suggested that productive coupling of the high-affinity agonist receptor complex to an *N*-ethylmaleimide-sensitive component, possibly adenylate cyclase, is also required for desensitization of the receptors in frog erythrocytes.

Recently Harden *et al.* (1979) have reported observations on a cultured cell line which appear to substantiate this formulation. Thus, in an astrocytoma cell line they found that, in addition to the fall in antagonist [^{125}I]iodo-hydroxybenzylpindolol binding to beta-adrenergic receptors that occurred during the desensitization process, there was an additional lesion termed an uncoupling of the receptor. Thus, early during the desensitization process, when catecholamine stimulation of adenylate cyclase was already significantly impaired, there was no fall in antagonist binding. However, it could be demonstrated that a displacement curve of [^{125}I]iodohydroxybenzylpindolol by an agonist, such as isoproterenol, was significantly shifted to the right. This is quite consistent with the inability to form the high-affinity state of the receptor described above in frog erythrocyte membranes. Harden *et al.* (1979) were able to show that uncoupling of the receptor appeared to precede temporally the actual loss of antagonist-binding sites. Thus, they speculated that there might be two distinct 'desensitized states' of the receptor which one might term R′ and R″. For example, R′ would be formed from the baseline state of the receptor under the influence of agonist and would represent the 'uncoupled state' of the receptor. R″ would represent another desensitized form of the receptor which would in turn arise from R′ and which would bind neither agonists nor antagonists. Such a formulation says nothing about the nature of these states. In fact, R″ might simply represent an absence of the receptor from the membrane, for example, if it had been lost from the membranes by a process such as internalization.

The situation in frog erythrocytes appears to be somewhat different. This system appears to be analogous to that described by Harden *et al.* (1979) in demonstrating the two distinct receptor alterations, that is, receptor loss and receptor uncoupling. However, in the frog erythrocyte we have been unable to demonstrate the temporal sequence observed by Harden *et al.* (1979) in the astrocytoma cells in which the uncoupling lesion preceded the apparent loss of receptor-binding capacity. In the frog erythrocyte system both lesions appear to be present even at early time-points. If in fact the scheme of Harden *et al.* (1979) is applicable to the frog erythrocyte model one might postulate that the rate of transition from R′ to R″ is much more rapid. If, for example, the transition from R to R′ were rapid but that from R′ to R″ slow, one might envisage a situation such as that observed by Harden *et al.* (1979); that is, R′ would build up, and one would measure a significant number of uncoupled receptors prior to the appearance of the actual loss of receptor binding capacity, i.e. R″. If instead the transition from R′ to R″ were rapid, then large quantities of R′ would not accumulate and one would, at any given point in time observe only a small amount of uncoupling (R′) and a more significant decrement in overall receptor binding capacity (R″). This situation might match more closely the frog erythrocyte system. Such formulations remain quite speculative.

At the present time firm evidence as to the fate of the desensitized beta-adrenergic receptors has not been published. In the *in vivo* frog erythrocyte system, as well as in the *in vitro* system, the process of desensitization and the loss of receptors both appear to be reversible (Mickey *et al.*, 1976; Mukherjee *et al.*, 1975, 1976). In the *in vivo* system the data suggest this could occur even without new protein synthesis (Mukherjee *et al.*, 1975, 1976). These findings might tend to suggest that the receptors are not actually lost from the membrane but rather that they might have been in some way inactivated by covalent or conformational change. However, no specific evidence about the nature of any such covalent or conformational changes has been presented, and no specific reagent or enzymatic treatment that will rapidly reverse the process in membranes has been discovered.

In analogy with a large number of polypeptide hormone receptors which appear to demonstrate the phenomenon of internationalization (Brown and Goldstein, 1979) it has been suggested that desensitized beta-adrenergic receptors may in fact be internalized and thereby lost from the membranes. Evidence in support of such a notion has been presented by Chuang and Costa (1979) using a bull-frog erythrocyte model. It was found that, coincident with desensitization and loss of beta-adrenergic receptor binding sites ([^{3}H]dihydroalprenolol-binding sites) from bull-frog erythrocyte particulate fractions, it could be demonstrated that there was an appearance of [^{3}H]dihydroalprenolol-binding sites in what was referred to as a 'soluble' fraction of the disrupted cells. In fact, this soluble fraction represented a 40 000 g supernatant which undoubtedly must have contained small membrane vesicles. [^{3}H]Dihydroalprenolol binding in this allegedly soluble preparation was studied by a Sephadex G-50 assay. Notwithstanding the fact that what these authors have referred to as a soluble receptor may have in fact represented receptors bound to very small membrane vesicles, these observations are still of interest since they suggest, at a minimum, a redistribution of the receptors from the normal plasma membrane fraction into smaller membrane vesicles during the desensitization process. This might well be compatible with an internalization mechanism. However, only a small fraction, perhaps 10 per cent or less, of the receptors lost from the plasma membranes could be accounted for in this supernatant fraction. The agonist-promoted appearance of the receptors in this supernatant fraction was blocked by antagonists as was the desensitization process. At the present time these data may be considered tantalizing and suggestive of the notion that desensitized beta-receptors may in fact be, at least to some extent, internalized. However, definitive proof of this hypothesis awaits further data.

While several studies have implicated alterations in hormone receptors as a significant mechanism in desensitization, in other systems alterations in hormone binding appear not to explain fully the decrease in hormone-stimulated adenylate cyclase activity. For example, Shear *et al*. (1976) found that desensitization of S49 lymphoma cells with isoproterenol resulted in a 60–80 per cent loss of isoproterenol-stimulated adenylate cyclase activity but only a 25–45 per cent reduction in binding of [^{125}I]iodohydroxybenzylpindolol. Johnson *et al*. (1978) reported an 80–90 per cent decrease in catecholamine-stimulated cAMP accumulation after incubation of astrocytoma cells with isoproterenol, but only a 20 per cent fall in beta-adrenergic receptor density, as assayed by [^{125}I]iodohydroxybenzylpindolol binding to the intact cells. As noted above, however, such discrepancies between the extent of diminution in beta-adrenergic antagonist binding and desensitization of adenylate cyclase may well be due to the superimposition of an 'uncoupling' lesion on the lesion that leads to an actual loss of receptors. However, they also appear to indicate that mechanisms involving components distal to the receptors, for example the nucleotide regulatory protein and the catalytic moiety of the enzyme, are also likely to be involved in certain desensitization phenomena. A reasonable hypothesis is that forms of catecholamine-induced desensitization which lead to loss of only beta-stimulated cyclase are primarily due to receptor alterations. By contrast, where more general alterations in enzyme activity are observed, it is probably the more distal components which are altered.

5.4 ADDITIONAL EXAMPLES OF HOMOLOGOUS DOWN-REGULATION

5.4.1 Leukocyte beta-adrenergic receptors

A very good system for studying physiologically relevant homologous down-regulation is the leukocyte system, since physiological and pharmacological data, as well as receptor binding data, are available. Beta-adrenergic stimulation of granulocytes inhibits both chemotaxis and lysosomal enzyme release; beta-adrenergic stimulation of lymphocytes inhibits cytolysis. These effects are mediated through cAMP accumulation intracellularly (Hatch *et al*., 1977). Using cAMP measurements, several groups have shown that leukocytes can be desensitized by *in vitro* exposure to the beta-agonist isoproterenol, with resultant fall in the isoproterenol-stimulated production of cAMP (Remold-O'Donnel, 1974; Lee, 1978). The beta-receptors on both granulocytes and lymphocytes have been identified directly by radioligand techniques (Williams *et al*., 1976; Davies and Lefkowitz, 1980b), and these receptors appear to mediate the isoproterenol stimulation of adenylate cyclase activity. Lee (1978) has shown that [^{3}H]DHA binding in lymphocytes falls after incubation with isoproterenol, and Conolly *et al*. (1976) showed a fall in cAMP production after similar incubation, consistent with the homologous desensitization observed in the adenylate cyclase.

Based on the animal and *in vitro* data on desensitization, one might expect homologous desensitization to occur in leukocytes from patients harboring pheochromocytomas, where endogenous catecholamine concentrations are elevated. Indeed, Greenacre and Conolly (1978) have demonstrated that lymphocytes from such patients were less able to produce cAMP in response to isoproterenol than those from control patients, with no alteration in lymphocytic phosphodiesterase. Similarly, low salt intake is known to increase plasma catecholamines in humans; Nadeau *et al.* (1980) have shown that these increases in catecholamines are associated with decreased human leukocytic beta-adrenergic receptor density. Physiologically, these changes in receptor density may represent important autoregulatory responses in those conditions in which plasma volume changes are hemodynamically important.

5.4.2 Asthma

Asthma is a human disease in which receptor regulation has been postulated. Szentivanyi (1968) proposed that the underlying atopic abnormality in asthma might be an imbalance in the alpha- and beta-adrenergically-mediated control mechanisms of bronchial reactivity. Several investigators have tested the hypothesis that there is a blockade of beta-receptor function in asthmatic patients by examining cAMP production in response to isoproterenol *in vitro*.

Parker and Smith (1973) demonstrated that lymphocytes from asthmatics had lower cAMP responses to isoproterenol as well as lower unstimulated cAMP production compared to normals, and this appeared to correlate in a general way with disease severity. Kalisker *et al.* (1977), Conolly and Greenacre (1976), and Sokol and Beall (1975), however, all found that isoproterenol-stimulated cAMP production in leukocytes was not different in asthmatics as compared with controls. However, adrenergic therapy (with bronchodilators) resulted in desensitization in the cAMP response to isoproterenol. To test the hypothesis of a beta-receptor abnormality in asthma further, others have examined the beta-receptors themselves with direct radioligand-binding studies in asthmatic patients on or off adrenergic therapy. Kariman (1980) reported less [3H]DHA binding in lymphocytes derived from asthmatic patients compared with those from controls. Galant *et al.* (1978) reported reduced [3H]DHA binding granulocytes after terbutaline treatment in both asthmatics and controls, but had too few subjects to determine reliably if a difference existed between untreated asthmatics and controls in [3H]DHA binding. Sano *et al.* (1979) supported both sets of observations by showing that lymphocytes from asthmatics had fewer receptor sites per cell than controls, whereas an equal number of sites per cell was found in granulocytes from asthmatics and controls. Also, they showed that both granulocytes and lymphocytes further lost receptor sites during chronic bronchodilator therapy. Certainly, the issues of beta-adrenergic receptor blockade or defect and of alpha-adrenergic hyperreactivity (Henderson *et al.*, 1979) in asthmatic patients is not yet resolved. However, Szentivanyi *et al.* (1979) demonstrated a fall in beta-adrenergic receptors and a rise in alpha-adrenergic receptors in both lung tissue and lymphocytes in persons with asthma.

5.4.3 Tricyclic antidepressants

Tricyclic antidepressants are thought to block the reuptake transport mechanisms at adrenergic synapses, thereby elevating the synaptic (extracellular) centration of neurotransmitters. One might expect elevation of norepinephrine at the postsynaptic receptors to result in homologous downregulation of the receptors. Indeed, administration of tricyclic antidepressants causes a fall in beta-receptor number in rat brains without an alteration in affinity for [^{3}H]DHA (Banerjee *et al.*, 1977a; Sarai *et al.*, 1978). Wolfe *et al.* (1978), using [^{125}I]HYP, obtained similar results and also showed a fall in isoproterenol-stimulated cAMP accumulation in such rat brains.

5.4.4 Cardiac hypertrophy

Cardiac hypertrophy induced by isoproterenol has been shown to decrease reversibly beta-adrenergic receptor density, as well as adenylate cyclase activity (Tse *et al.*, 1979). This may represent an isoproterenol effect on the receptors or it may represent a response of the receptors to hypertrophy, or both.

5.4.5 Ageing

The ageing process is associated with substantial alterations in susceptibility to cerebrovascular disease and in responsiveness to adrenergic therapy. In rat brains, the number of beta-receptors falls with ageing, and the extent to which light exposure causes a rise in beta-receptor number in pinealocytes (see below) is reduced as rats age (Roth, 1979; Greenberg and Weiss, 1978). That is, not only is the baseline receptor number lower, but the capacity to alter further that receptor number is reduced in aged animals. An interesting pattern of receptor regulation occurs in rat adipocytes during ageing: in the first eight months of life, beta-receptor number rises to about 40 800 sites per cell, but with senescence, the receptor number falls progressively to 7200 sites per cell (Giudicelli and Pecquery, 1978). There are parallel age-related alterations in catecholamine-sensitive adenylate cyclase. Similarly, Bylund *et al.* (1977) showed that as rats aged, a progressive decline in both beta-adrenergic receptors and adenylate cyclase was demonstrable on their erythrocytes.

In humans, a fall in responsiveness to infused isoproterenol occurs with ageing (Vestal *et al.*, 1979). Similarly, a fall in the production of cAMP in lymphocytes in response to isoproterenol occurs with ageing. Schocken and Roth (1977) showed a fall in human lymphocyte beta-receptors as the subjects aged. Since a gradual rise in plasma catecholamines occurs as humans age, it is probable that homologous desensitization is at least part of the explanation for the observed changes.

5.4.6 Alpha-adrenergic receptors

The alpha$_2$-receptors on human platelets also manifest homologous desensitization, suggesting that they may provide a useful model for studying receptor regulation in humans. It has been demonstrated that epinephrine-stimulated platelet aggregation is mediated through alpha-receptors. This effect occurs concurrently with an inhibition of the platelet adenylate cyclase. These alpha-receptors have been directly identified using [^{3}H]DHE (Alexander *et al.*, 1978; Newman *et al.*, 1978) and appear to be of the alpha$_2$-subtype (Hoffman *et al.*, 1979). Cooper *et al.* (1978) have demonstrated an apparent fall in alpha-receptor number after incubation of platelets with epinephrine. This reduction in receptor number is associated with a decreased ability of epinephrine to induce aggregation. Similarly, the alpha-adrenergic receptors on rat parotid cells undergo homologous down-regulation (Strittmatter *et al.*, 1977).

These examples demonstrate that homologous desensitization occurs for both the alpha- and beta-receptors, that these regulatory processes occur in man, as well as in experimental animals, and that they may be quite relevant to human disease or treatment states.

5.5 HOMOLOGOUS UP-REGULATION

5.5.1 Denervation hypersensitivity

The converse of down-regulation would be a rise in receptor density due to a fall in catecholamine concentrations. Such an up-regulation explains the well-known phenomenon of denervation hypersensitivity. Indeed, surgical interruption of innervation is known to increase beta-receptor density in rat skeletal muscle (Banerjee *et al.*, 1977b). Similarly, when lesions of the dorsal noradrenergic bundle are produced by local application of 6-hydroxydopamine, both alpha- and beta-adrenergic receptor density increase (U'Prichard *et al.*, 1980).

5.5.2 Pharmacological denervation

Up-regulation would be expected when a pharmacological agent reduces exposure of the receptor to catecholamines. 6-Hydroxydopamine destroys nerve terminals and depletes tissue catecholamines, and in rat brain, its administration is associated with an increase in beta-receptor sites (Sporn *et al.*, 1976) and is associated with a rise in isoproterenol-stimulated cAMP production. Further, apparently beta$_1$-adrenergic receptors participate in this regulation, but beta$_2$-adrenergic receptors do not (Minneman *et al.*, 1979). The up-regulatory effect of 6-hydroxydopamine also occurs in rat salivary gland alpha- and beta$_1$-adrenergic receptors (Pointon and Banjee, 1979), suggesting that this effect is quite general to several adrenergic receptor sub-types.

U'Prichard and Snyder (1978) reported that brains from reserpine-treated rats contained increased numbers of both alpha- and beta-receptors, as might be expected from the catecholamine depleting effect of reserpine. Guanethidine treatment might similarly be expected to result in catecholamine depletion; this is consistent with the finding of increased beta-receptors in rat hearts after chronic guanethidine treatment (Glaubiger *et al.*, 1978).

The interested reader is referred to the review of Molinoff *et al.* (1978) which contains a more detailed discussion of these phenomena.

5.5.3 Propranolol withdrawal syndrome

A very intriguing regulatory phenomenon induced by propranolol administration probably represents a unique example of indirect homologous up-regulation and will be discussed in this section. This phenomenon became of interest when Alderman *et al.* (1974) reported the occurrence of unstable angina pectoris or acute coronary events occurring during withdrawal from propranolol therapy. A composite postulate (Glaubiger and Lefkowitz, 1977; Boudoulas *et al.*, 1977; Davies and Lefkowitz, 1980a) to explain this phenomenon would suggest that this could represent a state of beta-adrenergic hypersensitivity, in which the presence of the antagonist, propranolol, effectively reduces the exposure of the receptors to catecholamines, thus leading to homologous up-regulation. However, presumably as the propranolol is withdrawn, the elevated receptor density persists, thus leaving excessive receptors exposed to catecholamines, and thus accounting for increased sensitivity. However, clear demonstration of a state of physiologically detectable hypersensitivity has been difficult. The evidence favoring such a hypersensitive state comes from two sources. First, Boudoulas *et al.* (1977) demonstrated that 24–48 hours after propranolol discontinuation, enhanced cardiac sensitivity to infused catecholamines existed in humans. Second, Nattel *et al.* (1979) demonstrated that of nine hypertensive human subjects, all demonstrated hypersensitivity to infused isoproterenol beginning two to six days after withdrawal of propranolol, and lasting for 3–13 days.

Studies from the receptor-binding point of view have also suggested that there may be alterations in the beta-receptors themselves after propranolol withdrawal. For example, Glaubiger and Lefkowitz (1977) showed a rise in beta-receptor number in rat myocardial membranes after two weeks of propranolol therapy. Dolphin *et al.* (1979) also demonstrated a 32 per cent rise in rat brain beta-receptors after propranolol treatment. More recently, Aarons *et al.* (1979) and Fraser *et al.* (1980) have demonstrated rises in beta-adrenergic receptor density in human leukocytes after proranolol administration. A major remaining issue, however, is to establish whether changes in receptor number after propranolol therapy are physiologically important in leading to the hypersensitivity to catecholamines in the so-called 'propranolol withdrawal syndrome'.

In these examples of up-regulation, a regulatory role of diminished exposure of alpha- and beta-adrenergic receptors to agonists is supported. Both receptor density and adenylate cyclase activity change in the same direction, although not always with the same magnitude. There appears to be a continuous range of catecholamine regulatory effects occurring such that as exposure to agonists increases from low levels to high levels, the receptor density falls from high values to low values.

5.6 HETEROLOGOUS REGULATION

We will discuss here examples of regulatory phenomena for which no mediation by catecholamines has been detected. However, it is realized that soon some of these phenomena may be sufficiently clearly investigated to justify their inclusion in another category of regulation. For clarity the regulatory phenomena are discussed according to the eliciting agent or condition, since a single agent or condition may induce heterologous regulation in different directions in different cells in a single model system.

5.6.1 Neoplasia

Adrenergic influences upon certain neoplasms may reflect altered receptor states. In cloned myeloid leukemic cells (Simantov and Sachs, 1978a) and in lymphocytic leukemic cells (Sheppard *et al.*, 1977), beta-receptor number is reduced. This may have functional significance since the myeloid series appears to contain at least two types of cells distinguishable by the ability of their adrenergic receptors to desensitize and by their ability to mature into granulocytes. MGI (macrophage and granulocyte inducer) is a protein that induces myeloid cells to become macrophages and polymorphonuclear leukocytes. One type of myeloid cell responds to this influence by differentiating; it has the capacity to down-regulate (reduce beta-receptor density) upon *in vitro* exposure to isoproterenol. Thus these cells are less sensitive to the cytotoxic effect of adrenergic stimulation. The second type of myeloid leukemic cell does not differentiate in response to MGI, its beta-adrenergic receptors do not desensitize in the presence of isoproterenol, and it is susceptible to adrenergic cytotoxicity. Thus, those malignant cells that are least controllable are also differentially more easily killed. These findings suggest that this may be an example of the fundamental utility of the desensitization process as a regulatory process which can help in the survival of a cell line in the face of a noxious stimulus (catecholamines in this case); moreover, there may be therapeutic implications from these findings. By understanding how desensitization occurs, the ability to impair desensitization could be developed, leading to the use of adrenergic agents as cytotoxic agents for some types of neoplasia.

Other neoplasms of interest in this context are certain types of adrenocortical carcinomas (Schorr *et al.*, 1971). Normal adrenal cortical cells do not bind [^{3}H]DHA well, but do bind ACTH with a concomitant rise in intracellular cAMP and subsequent steroidogenesis. Thus normal cells do not have beta-adrenergic receptors, but do contain functional ACTH receptors. In a particular rat adrenal cortical carcinoma, 'ectopic' [^{3}H]DHA binding sites (beta-receptors) are observed, and are coupled to adenylate cyclase. Therefore, in this model system, an adrenal carcinoma, steroidogenesis may result from adrenergic stimulation, providing an example of receptor-mediated ectopic hormone production in neoplasia (Williams *et al.*, 1977a).

The implications of these findings are that certain neoplasms may express their growth or humoral effects in a fashion dependent upon their adrenergic environment, even when one might not expect the corresponding normal tissues to exhibit much response to adrenergic stimuli.

5.6.2 Cardiovascular diseases

The widely known adrenergic actions upon cardiovascular tissues would naturally lead one to expect regulatory influences to be detected in cardiovascular diseases. Indeed, heterologous regulation occurs in response to at least three cardiovascular diseases.

Human hypertension results from many causes, but in only few is alpha-adrenergic hypersensitivity detectable. However, the effect of hypertension itself on the vessels might be expected to result in physiologically important regulatory phenomena. For example, it could be postulated that the arteries might protect themselves from excessive alpha-adrenergically mediated vasoconstriction with resultant further rise in intraluminal pressure by reducing the alpha-adrenergic receptor density, by increasing the beta-adrenergic receptor density, or by corresponding changes in the respective receptor's physiological effectiveness. However, thus far no reports have appeared which bear on alpha-adrenergic receptor binding in hypertensive humans or animals.

However, in the beta-adrenergic system, there is evidence in animals for an alteration of receptors. Cardiac (Limas and Limas, 1978; Woodcock *et al.*, 1978), renal (Woodcock *et al.*, 1978) and vascular (Limas and Limas, 1979) beta-receptors are diminished in spontaneously hypertensive rats. Such a down-regulation could prevent the counter-regulatory vasodilation that otherwise would result in maintenance of normal pressures. These changes would seem to reduce protective homeostatic mechanisms, and thus may represent primary phenomena rather than secondary autoregulatory changes.

In dog myocardium made ischemic by occlusion of the left anterior descending artery, beta-adrenergic receptor density rises after one hour and the elevation persists for at least eight hours (Mukherjee *et al.*, 1979). Apparently, this is a specific effect since muscarinic cholinergic receptor sites are not similarly altered. In another cardiac process, ventricular hypertrophy (induced by aortic banding), beta-adrenergic receptor density also rises (Limas, 1979). However, it should be noted as mentioned in Section 5.4.4 that when hypertrophy is generated by isoproterenol infusion, this effect is reversed. Further studies are certainly needed to establish the physiological relevance of receptor regulation in these important cardiovascular pathological processes.

5.6.3 Glucocorticoids

Glucocorticoid hormones have several important close interrelationships with the catecholamine hormones, a fact which is underscored by the anatomical relationship of the cells of origin of each type of hormone. Glucocorticoids potentiate glucose production in rat liver stimulated by catecholamines apparently through potentiation of the catecholamine-induced rise in cAMP production and in the responsiveness to the cAMP (Exton *et al.*, 1972). In human leukocytes, hydrocortisone itself apparently can stimulate adenylate cyclase activity (Logsdon *et al.*, 1972; Parker and Smith, 1973). In asthmatics this action and the potentiation of catecholamine stimulation can lead to a rise in leukocyte adenylate cyclase activity to a variable extent (Lee *et al.*, 1977). Methylprednisolone has been shown to protect dogs against desensitization of the isoproterenol-induced reduction in bronchoconstriction (Stephan *et al.*, 1980). Indeed, glucocorticoids can restore the lung beta-adrenergic receptors lost after adrenalectomy (Mano *et al.*, 1979) and can increase the rate of production of new beta receptors in cultured lung cells (Fraser and Venter, 1980). Leray *et al.* (1973) demonstrated that adrenalectomy led to the appearance of epinephrine-responsive adenylate cyclase in liver plasma membranes, when previously the enzyme had been unresponsive to catecholamines *in vitro* in the absence of glucagon. Chan *et al.* (1979) showed that adrenalectomy led to a rise in hepatic beta-adrenergic responses while alpha-adrenergic responses were diminished. Since adrenalectomy reduces glucocorticoid and catecholamine (at least initially) concentrations, this is consistent with either heterologous (steroid lack) or homologous (catecholamine lack) supersensitivity. Wolfe *et al*. (1976) correlated this rise in beta-adrenergic coupled adenylate cyclase with a three- to five-fold rise in beta-adrenergic receptor number in liver membranes prepared from adrenalectomized rats. This receptor number change could be reversed by cortisone replacement, but again whether this is a heterologous (directly cortisone-induced) or homologous [by steroid induction of phenylethanolamine-*N*-methyl transferase in paraganglional and Organ of Zuckerkandl sympathetic tissues (Wurtman and Axelrod, 1966)] change is not established. That the steroid induced change is specific for the beta-adrenergic receptors is supported by the finding of Guellaen *et al*. (1978) that normal rat liver membranes have 23-fold more alpha- than beta-adrenergic receptors, but that adrenalectomy caused no change in alpha-adrenergic receptor number while a twofold rise in beta-adrenergic receptor number occurred.

Cardiac effects of the catecholamines are also enhanced by corticosteroids. Guideri *et al*. (1978) determined that at a dose of $150\mu g$ of isoproterenol per kg body weight, 100 per cent of rats developed ventricular fibrillation in the presence of desoxycorticosterone acetate, while only 50 per cent of the rats developed the arrhythmia in the presence of prednisone. At this dose of isoproterenol, no steroid untreated rats developed the arrhythmia. This suggests a relative preponderance of the catecholamine-sensitizing effect with mineralocorticoid steroids. The inotropic effects may also be potentiated by steroids, as shown by Kaumann (1972) when he treated cat heart preparations with hydrocortisone. Isoproterenol-induced positive inotropism (but not norepinephrine-induced positive inotropism) was enhanced by the steroid, but when cocaine was used to block axonal reuptake, norepinephrine-induced positive inotropism was enhanced. Thus the apparent differential effect may relate best to reuptake mechanisms for norepinephrine, and indeed this suggests that both alpha- and beta-receptors in heart membranes may be altered by steroids. An intriguing footnote is the phenomenon of $ACTH_{1-24}$-enhanced positive inotropic effects of norepinephrine in an *in vitro* model in the absence of glucocorticoid (Basset *et al*., 1978). Hence myocardial inotropic and chronotropic effects of the pituitary–adrenal axis seem to be quite general phenomena with multiple chemically diverse agents being capable of inducing an effect.

Direct cardiac beta-adrenergic receptor measurements after adrenalectomy have given conflicting results. Abrass and Scarpace (1980) found a rise in receptor density. However, we found no change in receptor density post-adrenalectomy (Davies *et al*., 1981); the discrepancy presumably relates to differing experimental conditions. However, we did find a very intriguing form of receptor regulation in which the ability of the agonist, isoproterenol, to form the high-affinity activated form of the receptor was specifically diminished. This uncoupling of the receptor from the associated adenylate cyclase results in diminished physiological activity of the receptor and may represent a mechanism for fine-tuning the adrenergic responsiveness of a tissue.

Vascular adrenergic receptors may also be regulated by the presence or absence of steroid hormones. A very controversial relationship of steroids to catecholamines centers on the use of steroids in the treatment of septic shock. It may well be that a protective effect of steroids exists in humans as pointed out by Shumer (1976), and this effect may occur in part through regulation of the adrenergic receptors since their use is associated with an improvement of several cardiovascular parameters. Hydrocortisone, desoxycorticosterone, 17-beta-estradiol, and progesterone have all been shown *in vitro* to enhance catecholamine sensitivity of the aorta resulting in enhanced vasoconstriction which could raise arterial blood pressure (Besse and Bass, 1966; Kalsner, 1969). Altura and Altura (1974) further demonstrated that steroids block the catecholamine-induced contractions in precapillary arterioles thus enhancing microvascular circulation and therefore tissue perfusion. Indeed, the fact that only one exception to the rule (synephrine) that only catecholamine sympathomimetics are involved in this enhancement strongly suggests that the

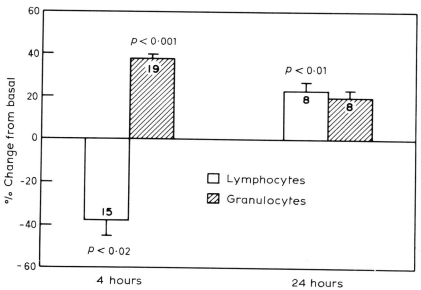

Fig. 5.5 Alteration in beta-adrenergic receptor density after cortisone administration. The ordinate represents the duration of treatment between pre- and post-treatment receptor density expressed as a per cent of the pretreatment value in fmol/mg protein. The bars shown are the SEM. The number of determinations is noted in the bars. From Davies and Lefkowitz (1980b).

regulation occurs at the level of the receptor where presumably the adrenergic specificity resides. Direct measurements of vascular adrenergic receptors will help to unravel the nature of the regulatory processes involved.

Finally, the only direct measurement of human beta-adrenergic receptor regulation induced by glucocorticoids thus far reported (Davies and Lefkowitz, 1980b) demonstrates a very important type of regulation: differential regulation. It was demonstrated that four hours after cortisone administration the beta-adrenergic receptor densities on granulocytes and lymphocytes were altered in opposing directions (see Fig. 5.5). Associated with the rise in granulocytic beta-adrenergic receptor density was a rise in adenylate cyclase activity. However, no change in lymphocytic adenylate cyclase activity was detected. These two cell types both contain receptors of the $beta_2$-subtype, yet they respond in opposite directions to the simultaneous exposure to a single agent. This 'differential regulation' suggests that regulation is dependent not only upon receptor subtype and the regulatory influence, but also upon local membrane or intracellular conditions.

5.6.4 Estrogens and progestogens

Estrogens and progestogens have regulatory effects on the contractile state of uterus, since the adrenergic response is dependent upon the steroidal environment. In both rabbits and humans estrogen treatment results in the enhancement of contraction (alpha), while pregnancy or progesterone treatment results in the enhancement of relaxation (beta) (Roberts *et al.*, 1977).

In concordance with these changes, Williams and Lefkowitz (1977a) indeed reported three-fold more alpha-adrenergic receptors in rabbits treated with estrogen compared with those treated with estrogen plus progesterone. Similar results were reported by Roberts *et al.* (1977) who also found no alteration in the beta-adrenergic receptors in these uteri. However, Krall *et al.* (1978) noted that in rats, estrogen treatment reduced the beta-adrenergic receptor number. Further, castration induced a rise in uterine beta-adrenergic receptor number. It is probably too simplistic to presume that receptor number changes alone explain the physiological alterations. This is due in part to the, as yet unclear, physiological significance of the $alpha_2$-receptor subtypes which can be detected by radioligand binding in uteri. It will be necessary to investigate further the direct regulatory role of sex steroids upon the receptors and their subtypes in order to understand these phenomena.

5.6.5 Hyperthyroidism

Hyperthyroidism represents a striking clinical constellation in which the lid lag and stare, tremor, tachycardia, and diaphoresis are manifestations consistent with a hyperadrenergic state. Levey (1971) approached the problem of the relationship of the hyperthyroid state to the adrenergic system by searching for evidence in the existing literature for enhanced cardiac sensitivity to infused catecholamines and in catecholamine-sensitive adenylate cyclase in dogs and rats that were hyperthyroid. He concluded that no such state could be shown to exist. However, Wildenthal (1974) found in an *in vitro* mouse heart culture system that incubation with tri-iodothyronine resulted in enhanced sensitivity to submaximal isoproterenol concentrations. Hashimoto and Nakashima (1978) later showed that thyroxine treatment of guinea pig atria or rabbit papillary muscle led to down-regulation of alpha-mediated positive inotropism while beta-mediated positive inotropism was enhanced. These observations leave somewhat open the question of hypersensitivity to catecholamines in the presence of hyperthyroidism, but suggest that a more sophisticated approach may clarify the existence of such a hypersensitivity.

Many groups have searched for alterations in receptor number in erythrocytes, hearts, adipocytes, and other tissues from hyperthyroid animals in an effort to define a hyperadrenergic state. Of these, the cardiac findings have been the most consistent. The existence of a state of hypersensitivity in cardiac tissues is supported by the finding that in a cultured heart system, the presence of tri-iodothyronine results in a rise in catecholamine-stimulated cAMP accumulation, and hearts from hyperthyroid rats produce more cAMP in response to isoproterenol (Tsai and Chen, 1978; Watanabe *et al.*, 1978; Tse *et al.*, 1980). Williams *et al.* (1977b) demonstrated a doubling of beta-receptor number in hyperthyroid rat heart membranes, with no change in affinity. Ciaraldi and Marinetti (1977) confirmed that hyperthyroid rat ventricular membranes have an increase in beta-adrenergic number. However, they further showed that the alpha-adrenergic receptor number was reduced, but with a decrease in apparent affinity for the antagonist, [3H]DHE. Similar increases in the cardiac beta-adrenergic receptor number have been reported by Tsai and Chen (1978) and Kempson *et al.* (1978). Williams and Lefkowitz (1979) have also reported a fall in cardiac alpha-adrenergic receptor binding in hyperthyroid rats, but this appeared to be predominantly due to a change in the affinity for the ligand without a clear reduction in receptor number. The situation with the alpha-adrenergic receptors is still unsettled, since changes in affinity for an antagonist have uncertain physiological significance.

Rather different situations have been described in other tissues. For example, turkey erythrocytes from hyperthyroid birds do not demonstrate a rise in beta-receptor number, but more cAMP was produced in response to low concentrations of isopreterenol than in normal erythrocytes (Bilezekian *et al.*, 1979). In rat adipocytes, hyperthyroidism induces enhancement of catecholamine-sensitive lipolysis, but no alteration in beta-receptor density or affinity could be detected (Malbon *et al.*, 1978; Goswami and Roseberg, 1978). Finally, in the only reported human study, beta-receptors on lymphocytes from hyperthyroid patients were unaltered, although a coupled biological response was not shown (Williams *et al.*, 1979). The extent to which human hyperthyroidism is associated with a change in receptors will remain uncertain until improved methods are utilized in several tissues.

5.6.6 Hypothyroidism

The relationship of hypothyroidism to adrenergic insensitivity has been noted in several species, including rats and humans. Rats normally undergo vasodilation, increased skin temperature, increased water intake, and increased plasma glucose in response to the beta agonist, isoproterenol. These responses are lost or blunted when the rats are made hypothyroid, and they return upon thyroxine replacement (Fregly *et al.*, 1975). Kunos *et al.* (1974) demonstrated that hypothyroid rats reduced their beta-mediated cardiac effects, while their alpha-mediated cardiac effects were enhanced. Radioligand-binding data actually demonstrated a fall in beta-adrenergic receptor number in rat heart, consistent with a heterologous down-regulation of the receptor by hypothyroidism (Ciaraldi and Marinetti, 1977; Banerjee and Kung, 1977).

In addition, Stiles *et al.* (1980) have demonstrated a fall in the ability of rat myocardial beta-adrenergic receptors after thyroidectomy to form the high-affinity complex HRN, similar to the post-adrenalectomy situation described above.

However, within the alpha-system the data are less clear. Williams and Lefkowitz (1979) reported no change in rat heart alpha-adrenergic receptors, but Ciaraldi and Marinetti (1977) report a fall in alpha-receptors while Sharma and Banerjee (1978b) report more alpha-adrenergic receptors in hearts from hypothyroid rats than in similar hearts after thyroxine treatment.

We have discussed here only the clearest findings relating thyroid pathology to adrenergic receptor regulation. A more detailed review by Williams and Lefkowitz (1980) is available to the interested reader. Some of the difficulties with the data may be resolved by improved techniques which can (1) quantitate changes in receptor subtypes or (2) quantitate the coupling of the receptor to adenylate cyclase.

5.7 SPECIAL CATEGORIES OF REGULATION

There are several forms of regulation which have been identified which are unique due to the nature of the interaction of the regulated receptor with the organism surrounding it. These we will describe in the context of the causative factors.

5.7.1 Phasic regulation

This might alternatively be termed cyclic or feedback regulation since there is some evidence that the existence of a primary regulatory receptor change can induce a secondary regulatory receptor change over a period of time. Examples of these intriguing forms of regulation follow.

(a) Neuronal tissues
The pinealocyte system is a good one in which to study phasic regulation since feedback neuroendocrinological input is substantial for these cells. Exposure of rats to light results in a general fall in sympathetic nervous activity through the action of the pineal gland. The fall in sympathetic activity might be expected to result in a rise in beta-receptor numbers. The pinealocytes themselves contain beta-receptors, and upon light exposure, their beta-receptor number indeed rises, as does their adenylate cyclase activity (Deguchi and Axelrod, 1973; Kebabian *et al.*, 1975). Conversely, upon administration of isoproterenol to rats, a fall in beta-receptor number is observed.

(b) Ethanol
The ethanol withdrawal state is clinically biphasic, with an early stage of florid signs and symptoms suggestive of a hyperadrenergic state, followed by a state of progressive resolution. This is consistent with the finding of reduced myocardial and cerebral beta-receptor numbers during ethanol-treatment of rats, but a substantial rise in beta-receptor number 48–72 hours after alcohol withdrawal (Banerjee *et al.*, 1978a). The elevation of receptor number occurs at a time consistent with the observed hyperadrenergic state, and may be contributory to it, thus representing an example of heterologous supersensitivity due to an increase in beta-receptor numbers. Of some interest in this regard is the report by Ciofalo (1978) that low concentrations of ethanol *in vitro* increase alpha-adrenergic receptor number in rat brains, while reducing receptor affinity, whereas higher ethanol concentrations reduce alpha-receptor number without altering affinity.

(c) Human leukocytes
As previously mentioned (see Section 5.5.3), cortisone given to humans induces altered beta-adrenergic receptor density after four hours. In that study it was also found that the receptors on lymphocytes respond differently at four hours (down-regulated) than at 24 hours (up-regulated), suggesting that in spite of continued cortisone administration, the lymphocyte receptors were responding to a secondary yet unidentified regulatory influence. In the case of isoproterenol infusions in humans, mononuclear leukocytes demonstrate an early (half hour) increment and then a late (four hour) decrement in beta-adrenergic receptor density (Tohmeh and Cryer, 1980). The precise mechanism of these phasic changes is unclear.

5.7.2 Cytoskeletal agents

There is accumulating evidence that the cytoskeleton contributes to certain of the receptor-regulatory processes under discussion. For example, when

Table 5.2 Additional miscellaneous examples of regulation of adrenergic receptor binding sites

Form	Type	Tissue	Direction	Reference
Homologous	Alpha	Rat parotid	Fall	Strittmatter et al. (1977)
Heterologous	Beta	Rat erythrocyte	Fall (ageing)	Bylund et al. (1977)
Heterologous	Beta	Rat adipocytes	Fall (cold)	Bukowiecki et al. (1978)
Heterologous	Beta	Rat submaxillary gland	Fall (thyroidectomy)	Pointon and Banerjee (1979)
Heterologous	Beta	Rat skeletal muscle	Fall (thyroidectomy)	Sharma and Banerjee (1978a)
Heterologous	Beta	Turkey erythrocyte	Fall (thyroidectomy)	Bilezekian et al. (1979)
Heterologous	Beta	Rat brain	Rise (pregnant mare serum)	Herndon et al. (1978)
Heterologous	Beta	He La cells	Rise (butyrate)	Tallman et al. (1977)

previously isoproterenol-desensitized leukemic cells are incubated with microtubule disrupting agents such as vinblastine, recovery of isoproterenol-stimulated adenylate cyclase occurs (Simantov and Sachs, 1978b). In non-desensitized S-49 lymphoma cells, colchicine augments isoproterenol-stimulated adenylate cyclase without altering agonist affinity (Insel and Kennedy, 1978). In contrast, in normal human leukocytes colchicine has been shown to increase agonist affinities (Rudolph *et al.*, 1977). These apparent discrepancies need to be more clearly examined, but it would appear that cytoskeletal elements have regulatory influences, perhaps by altering the interactions of the several components, R, N, and C, necessary for receptor function.

5.7.3 Mixed or uncertain regulatory processes

Examples of regulatory processes that are mixed, uncertain, or otherwise not clearly part of other regulation categories include the finding that diabetic rat hearts contain fewer beta-adrenergic receptors than controls (Savarese and Berkowitz, 1979). Another example is that of amphetamine administration. In analogy with the phenomenon of indirect heterologous up-regulation, acute amphetamine administration through depleting catecholamine stores results in rat brain membranes containing increased numbers of beta-adrenergic receptors (Banerjee *et al.*, 1978b; Howlett and Nahorski, 1978). By contrast, after chronic amphetamine exposure, a reduction in isoproterenol-stimulated cAMP accumulation and in beta-adrenergic receptors occurs. This may be due to the postulated stimulatory effect of amphetamine on beta-adrenergic receptors, in analogy with direct homologous desensitization. Several other examples of mixed regulation are noted in Table 5.2.

5.8 IMPLICATIONS

The earliest experimental approaches to the direct study of adrenergic receptor regulation were simply measurements of the numbers of receptors under various experimental conditions. Often changes in receptor number were observed. However, when it became apparent that expected changes in receptor density did not always occur, questions about receptor affinity were raised. A major advance in receptor regulation studies occurred when the biochemical steps in receptor function began to be elucidated. It has recently become possible to observe regulatory effects upon some of the more subtle biochemical steps in receptor activation. It would appear that there are many steps in the pathway of intercellular adrenergic communication from the release of the mediator to its receptor binding, through the processes of receptor activation to the final cellular response. For every step along the pathway, one might expect eventually to find at least one agent or condition which exerts its regulatory effect at that step.

Summarized in this chapter have been examples of regulation in which receptor number or affinity change, where the receptor's ability to form a high-affinity 'activated' state of the receptor is altered, or where the associated adenylate cyclase enzyme is altered. These changes may stem directly from the agent or condition in question, or may be more indirect. Indeed, they may occur in response to changes in other receptor systems, or to cytoskeletal alterations, or to as yet unknown indirect causes. The changes may be primary or they may be autoregulatory in nature. The result of investigations of adrenergic receptor regulation should be a greater understanding of the biochemical processes involved in receptor function, and potentially the modification of these processes for therapeutic gain.

REFERENCES

General Reviews

Davies, A.O and Lefkowitz, R.J. (1980a), *Recent Advances in Clinical Pharmacology*, **2**, 35–54.

Hoffman, B.B. and Lefkowitz, R.J. (1980), *Annu. Rev. Pharm. Tox.*, **20**, 581–608.

Lefkowitz, R.J., Limbird, L. E., Mukherjee, C. and Caron, M. G. (1976), *Biochim. Biophys. Acta*, **457**, 1–39.

Williams, L.T. and Lefkowitz, R.J. (1978), *Receptor Binding Studies in Adrenergic Pharmacology*, Raven Press, New York.

Wolfe, B.B., Harken, T.K. and Molinoff, P.B. (1977), *Annu. Rev. Pharm. Tox.*, **17**, 575–604.

Other References

Aarons, R.D., Nies, A.S., Gal, J. and Molinoff, P.B. (1979), *Proc. West. Pharm. Soc.*, **22**, 175–176.

Abrass, I.B. and Scarpace, P.J. (1980), *Fed. Proc.*, **37**, 1788 (abstract 2843).

Ahlquist, R.P. (1948), *Am. J. Phys.*, **153**, 586–600.

Alderman, E.L., Coltart, D.J., Wettach, G.E. and Harrison, D.C. (1974), *Ann. Int. Med.*, **81**, 625–627.

Alexander, R.W., Cooper, B. and Handin, R.I. (1978), *J. Clin. Invest.*, **61**, 1136–1155.

Altura, B.M. and Altura, B.T. (1974), *J. Pharm. Exp. Ther.*, **190**, 300–315.

Aurbach, G.D., Fedak, S.A., Woodward, C.J., Palmer, J.S., Hauser, D. and Troxler, F. (1974), *Science*, **186**, 1223–1224.

Banerjee, S.P. and Kung, L.S. (1977), *Eur. J. Pharm.*, **43**, 207–208.

Banerjee, S.P., Kung, L.S., Riggi, S.J. and Chanda, S.K. (1977a), *Nature*, **268**, 454–456.

Banerjee, S.P., Sharma, V.K. and Khanna, J.M. (1978a), *Nature*, **276**, 407–409.

Banerjee, S.P., Sharma, V.K. and Kung, L.S. (1977b), *Biochim. Biophys.* Acta, **470**, 123–127.

Banerjee, S.P., Sharma, V.K., Kung, L.S. and Chanda, S.K. (1978b), *Nature*, **271**, 380–381.

Basset, J.R., Strand, F.L. and Cairncross, K.D. (1978), *Eur. J. Pharm.*, **49**, 243–249.

Berthelson, S. and Pettinger, W.A. (1977), *Life Sci.*, **2**, 595–606.

Besse, J.C. and Bass, A.D. (1966), *J. Pharm. Exp. Ther.*, **154**, 224–238.

Bilezekian, J.P., Loeb, J.N. and Gammon, D.E. (1979), *J. Clin. Invest.*, **63**, 184–192.

Boudoulas, H., Lewis, R.P., Kates, R.E. and Dalamangas, G. (1977), *Ann. Int. Med.*, **87**, 433–436.

Brown, M.S. and Goldstein, J.L. (1979), *Proc. Natl. Acad. Sci.*, *USA*, **76**, 3330–3337.

Bukowiecki, L., Follea, N., Vallieres, J. and Lebalanc, J. (1978), *Eur. J. Biochem.*, **92**, 189–196.

Bylund, D.B., Tellez-Inon, M.T. and Hollenberg, M.D. (1977), *Life Sci.*, **21**, 403–410.

Caron, M.G., Srinivasan, Y., Pitha, J., Kociolek, K. and Lefkowitz, R.J. (1979), *J. Biol. Chem.*, **254**, 2923–2927.

Chan, T.M., Blackmore, P.F., Steines, K.E. and Exton, J.H. (1979), *J. Biol. Chem.*, **254**, 2428–2433.

Chuang, D.M. and Costa, E. (1979), *Proc. Natl. Acad. Sci. USA*, **76**, 3024–3028.

Ciaraldi, T. and Marinetti, G.V. (1977), *Biochem. Biophys. Res. Commun.*, **74**, 984–991.

Ciofalo, F.R. (1978), *Proc. West. Pharm. Soc.*, **21**, 267–269.

Conolly, M.E., Greenacre, J.K. and Scofield, P. (1976), *Br. J. Pharm.*, **58**, 448P–449P.

Conolly, M.E. and Greenacre, J.K. (1976), *J. Clin. Invest.*, **58**, 1307–1316.

Cooper, B., Handin, R.I., Young, L.H. and Alexander, R.W. (1978), *Nature*, **274**, 703–706.

Davies, A.O. and Lefkowitz, R.J. (1980b), *J. Clin. End. Metab.*, **51**, 599–605.

Davies, A.O., DeLean, A. and Lefkowitz, R.J. (1981), *Endocrinology*, **108**, 720–722.

Deguchi, T. and Axelrod, J. (1973, *Proc. Natl. Acad. Sci. USA*, **70**, 2411–2414.

De Lean, A., Munson, P.J. and Rodbard, D. (1978), *Am. J. Phys.*, **235**, E97–E102.

De Lean, A., Stadel, J.M. and Lefkowitz, R.J. (1980), *J. Biol. Chem.*, **255**, 7108–7117.

Dolphin, A., Adrien, J., Hamon, M. and Bockaert, J. (1979), *Mol. Pharm.*, **15**, 1–15.

Exton, J.H., Friedmann, N., Wong, E. H-A., Brineaux, J.P., Corbin, J.D. and Park, C.R. (1972), *J. Biol. Chem.*, **247**, 3579–3588.

Fraser, C.M. and Venter, J.C. (1980), *Biochem. Biophys. Res. Commun.*, **94**, 390–397.

Fraser, J., Nadeau, J.H., Robertson, D. and Wood, A.J.J. (1980), *Clin. Res.*, **28**, 541A.

Fregly, M.J., Nelson, E.L. Jr., Resch, G.E., Field, F.P. and Lutherer, L.O. (1975), *Am. J. Phys.*, **229**, 916–924.

Galant, S.P., Duriseti, L., Underwood, S. and Insel, P.A. (1978), *N.Engl. J. Med.*, **299**, 933–936.

Giudicelli, Y. and Pecquery, R. (1978), *Eur. J. Biochem.*, **90**, 413–419.

Glaubiger, G. and Lefkowitz, R.J. (1977), *Biochem. Biophys. Res. Commun.*, **78**, 720–725.

Glaubiger, G., Tsai, B.S. and Lefkowitz, R.J. (1978), *Nature*, **273**, 240–242.

Goswami, A. and Rosenberg, I. (1978), *Endocrinology*, **103**, 2223–2233.

Greenacre, J.K. and Conolly, M.E. (1978), *Eur. J. Clin. Pharm.*, **5**, 191–197.

Greenberg, L.H. and Weiss, B. (1978), *Science*, **201**, 61–63.

Guellaen, G., Yates-Aggerbeck, M., Vauquelin, G., Strosberg, D. and Hanoune, J. (1978), *J. Biol. Chem.*, **253**, 1114–1120.

Guideri, G., Green, M. and Lehr, D. (1978), *Res. Commun. Chem. Path. Pharm.*, **21**, 197–212.

Hancock, A.A., DeLean, A.L. and Lefkowitz, R.J. (1979), *Mol. Pharm.*, **16**, 1–9.

Harden, T.K., Su, Y.F. and Perkins, J.P. (1979), *J. Cyc. Nuc. Res.*, **5**, 99–106.

Hashimoto, H. and Nakashima, M. (1978), *Eur. J. Pharm.*, **50**, 337–347.

Hatch, G.E., Nichols, W.K. and Hill, H.R. (1977), *J. Immunol.*, **119**, 450–456.

Henderson, W.R., Shelhamer, J.H., Reingold, D.B., Smith, W., Evans, III, R. and Kaliner, M. (1979), *N. Engl. J. Med.*, **300**, 642–647.

Herndon, H., Wilkinson, M., Wilson, C.A. (1978), *J. Physiol. (London)*, **276**, 63P–64P.

Hoffman, B.B., DeLean, A., Wood, C.L., Shocken, D.D. and Lefkowitz, R.J. (1979), *Life Sci.*, **24**, 1739–1746.

Hoffman, B.B., Mullikin-Kilpatrick, D. and Lefkowitz, R.J. (1980), *J. Biol. Chem.*, **255**, 4645–4652.

Howlett, D.R. and Nahorski, S.R. (1978), *Br. J. Pharm.*, **64**, 411–412P.

Insel, P.A. and Kennedy, M.S. (1978), *Nature*, **273**, 471–473.

Insel, P.A., Nirenberg, P., Turnbull, J. and Shattil, S.J. (1978), *Biochemistry*, **17**, 5269–5274.

Johnson, G.L., Wolfe, B.B., Harden, T.K., Molinoff, P.B. and Perkins, J.P. (1978), *J. Biol. Chem.*, **253**, 1472–1480.

Kalisker, A., Nelson, H. E. and Middleton, E. Jr. (1977), *J. All. Clin. Immunol.*, **60**, 259–265.

Kalsner, S. (1969), *Br. J. Pharm.*, **36**, 582–593.

Kariman, K. (1980), *Lung*, **157**, 1–11.

Kaumann, A.J. (1972), *Naun. Schmied. Arch. Pharm.*, **273**, 134–153.

Kebabian, J.W., Zatz, M., Romero, J.A. and Axelrod, J. (1975), *Proc. Natl. Acad. Sci. USA*, **72**, 3735–3739.

Kempson, S., Marinetti, G.V. and Shaw, A. (1978), *Biochim. Biophys. Acta*, **540**, 320–329.

Kent, R., De Lean, A. and Lefkowitz, R.J. (1980), *Mol. Pharmacol.*, **17**, 14–23.

Krall, J.F., Mori, H., Tuck, M.L., LeShan, S.L. and Koreman, S.G. (1978), *Life Sci.*, **23**, 1073–1082.

Kunos, G., Vermes-Kunos, I. and Niederson, M. (1974), *Nature*, **250**, 779–781.

Lands, A.M., Arnold, A., McAuliff, J.P., Luduena, F.P. and Brown, T.G. (1967), *Nature*, **214**, 597–598.

Langer, S.Z. (1974), *Biochem, Pharmacol.*, **23**, 1793–1800.

Lee, T.P. (1978), *Res. Commun. Chem. Path. Pharm.*, **22**, 223–242.

Lee, T.P., Basse, W.W. and Reed, C.E. (1977), *J. All. Clin. Immun.*, **59**, 408–413.

Lefkowitz, R.J., Mukherjee, C., Coverstone, M. and Caron, M.G. (1974), *Biochem. Biophys. Res. Commun.*, **60**, 703–709.

Lefkowitz, R.J., Mukherjee, C., Limbird, L.E., Caron, M.G., Williams, L.T.,

Alexander, T.W., Mickey, J.V. and Tate, R. (1976), *Rec. Prog. Horm. Res.*, **32**, 597–632.

Lefkowitz, R.J., Mullikin, D., Wood, C.L., Gore, T.B. and Mukherjee, C.J. (1977), *J. Biol. Chem.*, **252**, 5295–5303.

Leray, F., Chambaut, A.M., Perrenaud, M.L. and Hanoune, J. (1973), *Eur. J. Biochem.*, **38**, 185–192.

Levey, G.S. (1971), *Am. J. Med.*, **50**, 413–420.

Levitski, A., Atlas, D. and Steer, M.L., (1974), *Proc. Natl. Acad. Sci. USA*, **71**, 2773–2776.

Limas, C. and Limas, C.J. (1978), *Biochem. Biophys. Res. Commun.*, **83**, 710–714.

Limas, C.J. (1979), *Biochim. Biophys. Acta*, **588**, 174–178.

Limas, C.J. and Limas, C. (1979), *Biochim. Biophys. Acta*, **582**, 533–535.

Limbird, L.E., Gill, D.M. and Lefkowitz, R.J. (1980), *J. Biol. Chem.*, **255**, 1854–1861.

Limbird, L.E. and Lefkowitz, R.J. (1978), *Proc. Natl. Acad. Sci, USA*, **75**, 228–232.

Logsdon, P.J., Middleton, E. and Coffey, R.G. (1972), *J. All. Clin. Immun.*, **50**, 45–56.

Malbon, C.C., Moreno, F.J., Cabelli, R.J. and Fain, J.N. (1978), *J. Biol. Chem.*, **253**, 671–678.

Mano, K., Akbarzadeh, A. and Townley, R.G. (1979), *Life Sci.*, **25**, 1925–1930.

Mickey, J.V., Tate, R. and Lefkowitz, R.J. (1975), *J. Biol. Chem.*, **250**, 5727–5729.

Mickey, J.V., Tate, R., Mullikin, D. and Lefkowitz, R.J. (1976), *Mol. Pharacol.*, **12**, 409–419.

Minneman, K.P., Dibner, M.D., Wolfe, B.B. and Molinoff, P.B. (1979), *Science*, **209**, 866–868.

Molinoff, P.B., Sporn, J.R., Wolfe, B.B. and Harden, T.K. (1978), *Adv. Cyc. Nuc. Res.*, **9**, 465–483.

Mukherjee, C., Caron, M.G. and Lefkowitz, R.J. (1975), *Proc. Natl. Acad. Sci. USA*, **72**, 1945–1949.

Mukherjee, C., Caron, M.G. and Lefkowitz, R.J. (1976), *Endocrinology*, **99**, 347–357.

Mukherjee, A., Wong, T.M., Buja, L.M., Lefkowitz, R.J. and Willerson, J.T. (1979), *J. Clin. Invest.*, **64**, 1423–1428.

Nadeau, J.H., Fraser, J., Robertson, D. and Wood, A.J.J. (1980), *Clin. Res.*, **28**, 240A.

Nattel, S., Rango, R.E. and Van Loon, G. (1979), *Circ.*, **59**, 1158–1164.

Newman, K.D., Williams, L.T., Bishopric, N.H. and Lefkowitz, R.J. (1978), *J. Clin. Invest.*, **61**, 395–402.

Parker, C.W. and Smith, J.W. (1973), *J. Clin. Invest.*, **52**, 48–59.

Pointon, S.E. and Banerjee, S.P. (1979), *Biochim. Biophys. Acta*, **583**, 129–132.

Remold-O'Donnell, E. (1974), *J. Biol. Chem.*, **249**, 3615–3621.

Roberts, J.M., Insel, P.A., Goldfien, R.D. and Goldfien, A. (1977), *Nature*, **270**, 624–625.

Roth, G.S. (1979), *Fed. Proc.*, **38**, 1910–1914.

Rudolph, S.A., Greengard, P. and Malawista, S.E. (1977), *Proc. Natl. Acad. Sci. USA*, **74**, 3404–3408.

Sano, Y., Ruprecht, H., Mano, K., Begley, M., Bewtra, A. and Townley, R. (1979), *Clin. Res.*, **27**, 403A.

Sarai, K., Frazer, A., Brunswich, D. and Mendels, J. (1978), *Biochem. Pharmacol.*, **27**, 2179–2181.

Savarese, J.J. and Berkowitz, B.A. (1979), *Life Sci.*, **25**, 2075–2078.

Schocken, D.D. and Roth, G.S. (1977), *Nature*, **267**, 85–87.

Schorr, I., Rathman, P., Saxena, B.B. and Ney, R.L. (1971), *J. Biol. Chem.*, **246**, 5806–5811.

Sharma, V.K. and Banerjee, S.P. (1978a), *Biochim. Biophys. Acta*, **539**, 538–542.

Sharma, V.K. and Banerjee, S.P. (1978b), *J. Biol. Chem.*, **253**, 5277–5279.

Shear, M., Insel, P.A., Melmon, K.L. and Coffino, P.J. (1976), *J. Biol. Chem.*, **251**, 7572–7576.

Sheppard, J.R., Gormus, R. and Molden, C.F. (1977), *Nature*, **269**, 693–695.

Shumer, W. (1976), *Ann. Surg.*, **184**, 333–339.

Simantov, R. and Sachs, L. (1978a), *Proc. Natl. Acad. Sci. USA*, **75**, 1805–1809.

Simantov, R. and Sachs, L. (1978b), *FEBS Lett.*, **90**, 69–75.

Sokol, W.N. and Beall, G.N. (1975), *J. All. Clin. Immun.*, **55**, 310–324.

Sporn, J.R., Harden, T.K., Wolfe, B.B. and Molinoff, P.B. (1976), *Sci.*, **194**, 624–625.

Stadel, J.M. and Lefkowitz, R.J. (1979), *Mol. Pharmacol.*, **16**, 709–718.

Stadel, J.M. De Lean, A. and Lefkowitz, R.J. (1981), *Adv. Enzymol.*, in press.

Starke, K., Endo, T. and Taube, H.D. (1975), *Naunyn-Schmied. Arch. Pharm.*, **291**, 55–78.

Stephan, W.C., Chick, T.W., Avner, B.P. and Jenne, J.W. (1980), *J. All. Clin. Immun.*, **65**, 105–109.

Stiles, G.L., De Lean, A. and Lefkowitz, R.J. (1981), submitted for publication.

Strittmatter, W.J., Davis, J.N. and Lefkowitz, R.J. (1977), *J. Biol. Chem.*, **252**, 5478–5482.

Sutherland, E.W. (1971), in *Cyclic AMP*, (Robison, G.A., Butcher, R.W. and Sutherland, E.W., eds), Academic Press, New York, pp. 1–6.

Szentivanyi, A. (1968), *J. Allergy*, **42**, 203–232.

Szentivanyi, A., Heim, O. and Schultze, P. (1979), *Ann. N.Y. Acad. Sci.*, **332**, 295–298.

Tallman, J.F., Smith, C.C. and Henneberry, R.C. (1977), *Proc. Natl. Acad. Sci. USA*, **74**, 873–877.

Tohmeh, J.F. and Cryer, P.E. (1980), *J. Clin. Invest.*, **65**, 836–840.

Tsai, J.S. and Chen, A. (1978), *Nature*, **275**, 138–140.

Tse, J., Powell, J.R., Baste, C.A., Priest, R.E. and Kuo, J.F. (1979), *Endocrinology*, **105**, 246–255.

Tse, J., Wrenn, R.W. and Kuo, J.F. (1980), *Endocrinology*, **107**, 6–16.

U'Pritchard, D.C., Reisine, T.D., Mason, S.T., Fibiger, H.C. and Yamamura, H.I. (1980), *Brain Res.*, **187**, 143–154.

U'Pritchard, D.C. and Snyder, S.H. (1978), *Eur. J. Pharm.*, **51**, 145–155.

Vestal, R.E., Wood, A.J.J. and Shand, D.G. (1979), *Clin. Pharm. Ther.*, **26** 181–186.

Watanabe, A.M., Hathaway, D.R. and Besch, H.R. (1978), *Rec. Adv. Stud. Card. Struc. Metab.*, **11**, 423–429.

Wessels, M.R., Mullikin, D. and Lefkowitz, R.J. (1978), *J. Biol. Chem.*, **253**, 3371–3373.

Wessels, M.R., Mullikin, D. and Lefkowitz, R.J. (1979), *Mol. Pharmacol.*, **16**, 10–20.

Wildenthal, K. (1974), *J. Pharm. Exp. Ther.*, **190**, 277–279.

Williams, L.T., Gore, T.B. and Lefkowitz, R.J. (1977a), *J. Clin. Invest.*, **59**, 319–324.

Williams, L.T. and Lefkowitz, R.J. (1977a), *J. Clin. Invest.*, **60**, 815–818.

Williams, L.T. and Lefkowitz, R.J. (1977b), *J. Biol. Chem.*, **252**, 7207–7213.

Williams, L.T., Lefkowitz, R.J., Watanabe, A.M., Hathaway, D.R. and Besch, H.R. (1977b), *J. Biol. Chem.*, **252**, 2787–2789.

Williams, L.T., Snyderman, R. and Lefkowitz, R.J. (1976), *J. Clin. Invest.*, **57**, 149–155.

Williams, R.S., Guthrow, C.E. and Lefkowitz, R.J. (1979), *J. Clin. Endo. Metab.*, **48**, 503–505.

Williams, R.S. and Lefkowitz, R.J. (1979), *J. Card. Pharm.*, **1**, 181–189.

Williams R.S. and Lefkowitz, R.J. (1980), In *Molecular Basis of Thyroid Hormone Action*, in press.

Wolfe, B.B., Harden, T.K. and Molinoff, P.B. (1976), *Proc. Natl. Acad. Sci. USA*, **73**, 1343–1347.

Wolfe, B.B., Harden, T.K., Sporn, J.R. and Molinoff, P.B. (1978), *J. Pharm. Exp. Ther.*, **207**, 446–457.

Woodcock, E.A., Funder, J.W. and Johnston, C.I. (1978), *Clin. Exp. Pharm. Phys.*, **5**, 545–550.

Wurtman, R.J. and Axelrod, J. (1966), *J. Biol. Chem.*, **241**, 2301–2305.

6 Denervation: Cholinergic Receptors of Skeletal Muscle

DOUGLAS M. FAMBROUGH

Acknowledgement

Research in the author's laboratory has been supported in part by a grant from the Muscular Dystrophy Association.

Receptor Regulation
(*Receptors and Recognition*, Series B, Volume 13)
Edited by R. J. Lefkowitz
Published in 1981 by Chapman and Hall, 11 New Fetter Lane, London EC4P 4EE
© 1981 Chapman and Hall

Denervation supersensitivity is recognized as a fairly general phenomenon in denervated target organs, recognized to the extent that the appearance of denervation supersensitivity in any denervated tissue is considered in keeping with Cannon's 'Law' (see Cannon and Rosenblueth, 1949). With no *a priori* reason to expect it, physiologists have found that denervated tissues responded to the appropriate neurotransmitter with heightened sensitivity. Do the same molecular mechanisms underly the development of denervation supersensitivity in various tissues, as Occam's razor would bias one to presume? Probably not. The 'sensitivity' of a tissue to neurotransmitters, we now know, depends upon several variables, including some variables which are pertinent only to a few target tissues. For example, the action of some transmitters is terminated by presynaptic transmitter-uptake mechanisms which are destroyed by denervation, rendering the response to a pulse of exogenous transmitter more prolonged and the effective concentration of transmitter at the receptor sites higher than in innervated tissues.

At the vertebrate neuromuscular junction, which is a cholinergic synapse, the action of acetylcholine is terminated largely by hydrolysis via the enzyme acetylcholinesterase, which is associated with the synaptic cleft material and remains enzymatically effective for some time after denervation. If loss of acetylcholinesterase activity is in part responsible for denervation supersensitivity of denervated skeletal muscle, its contributions must be modest (Pecot-Dechavassine, 1968). Several other denervation changes contribute modestly to alterations in sensitivity to acetylcholine. The specific resistance of the plasma membrane increases about twofold (Albuquerque and McIsaac, 1970) and the input resistance of the muscle fibers increases due to this and also due to the decrease in muscle fiber diameter. The open time of the ACh receptor ion channels is somewhat longer for most of the ACh receptors present in denervated muscle (Neher and Sakmann, 1976), which may contribute to a larger summed response to acetylcholine. Counterbalancing these changes, there is a fall in muscle fiber resting membrane potential and a decrease in unit conductance of ACh receptor-associated ion channels. The overall effect of these changes on denervation supersensitivity probably varies with time after denervation and is likely to vary also between species and even between muscles within a given species. (The relative importance of these changes also depends upon the method by which denervation supersensitivity is measured.) All of these contributions are small, however, compared with the change in sensitivity due to the post-denervation appearance of a very large number of ACh receptors. The change in the number and distribution of acetylcholine receptors following denervation is the focus of this chapter.

The increased chemosensitivity of denervated skeletal muscle to acetylcholine is due to an increase in sensitivity of extrajunctional regions of muscle (Axelsson and Thesleff, 1959). The standard, semi-quantitative way of measuring this increase has been to determine the relative amount of acetylcholine needed to depolarize the muscle fiber membrane by 1 mV (Miledi, 1960). The method requires that a recording microelectrode be placed in a muscle fiber and extracellular micropipette containing acetylcholine be manipulated up to the muscle membrane and acetylcholine ejected in a metered way by electrophoresis. Then the position of the micropipette and the strength and duration of the electrophoretic pulse of acetylcholine are varied until the maximum response per unit dose is achieved. Typically this might involve a 2 millisecond pulse, delivering a total current of 10^{-11} coulombs of charge (part of which is carried by acetylcholine cations), and resulting in a local 1 mV depolarization of the muscle fiber. Positioned along the muscle fiber, the recording electrode and ACh micropipette can be used to 'map' the chemosensitivity. When muscle fibers are mapped at various times after denervation, a fairly concerted increase in chemosensitivity is seen along the muscle fiber (Fig. 6.1) with little change in chemosensitivity at the position of the former neuromuscular junction.

The numbers of ACh receptor sites at neuromuscular junctions and in extrasynaptic regions have been quantified, using radioactive derivatives of

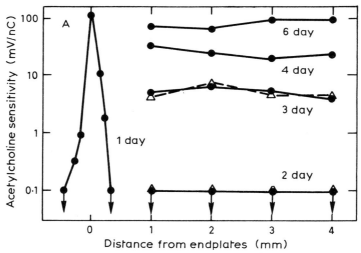

Fig. 6.1 Acetylcholine sensitivity of rat diaphragm muscle fibers 1, 2, 3, 4, and 6 days after section of the phrenic nerve ($\bullet$) or 2 and 3 days after transfer of muscle to organ culture ($\triangle$) (a process which involves denervation). The arrows indicate that the sensitivity was less than 0.1 mV/nC and usually below the limit of detection. (From Fambrough, 1970.)

α-bungarotoxin. This toxin, which is a polypeptide of molecular weight about 8000 daltons, is isolated from the venom of the Formosan banded krait, *Bungarus multicinctus* (Lee, 1972). High toxicity is retained during derivatizations such as iodination, acetylation and fluorescein conjugation. The number of toxin binding sites per neuromuscular junction is the range $2–5 \times 10^7$ in various muscles of man, mouse, rat, chicken and frog. Electron microscope radioautographic studies have established that the receptor sites are localized in the juxtaneural portions of the post-synaptic membrane, occupying an area which is subtended by cytoplasmic material which appears electron-dense after heavy-metal staining (Fertuck and Salpeter, 1974). In these areas the packing density of receptor sites is about 20 000 per μm^2 of membrane. This packing density and the total number of receptors in these areas do not change very much following denervation. On the other hand, the packing density of receptors in extrajunctional plasma membrane rises from a few (or no) sites per square micrometer to several hundred (Fambrough, 1974). The time course for such a change following denervation of rat diaphragm muscle is illustrated in Fig. 6.2, and example radioautographs are shown in Fig. 6.3.

In the case of the rat diaphragm, the maximum number of receptors occur-

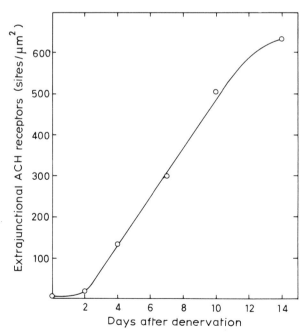

Fig. 6.2 Accumulation of extrajunctional ACh receptors in rat diaphragm muscle after denervation, measured as ^{125}I-α-bungarotoxin-binding sites per unit area of extrajunctional surface. (Drawn from data of Fambrough, 1974.)

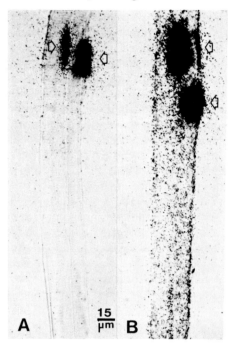

Fig. 6.3 Radioautographs of pairs of muscle fibers teased from normal (A) and denervated (B) rat diaphragms after ACh receptors were labelled with ^{125}I-α-bungarotoxin. Note clusters of silver grains in each panel, indicating the positions of the neuromuscular junctions or residual post-synaptic element thereof (arrows). The appearance of extrajunctional ACh sensitivity in the denervated muscle does not involve a redistribution of synaptic receptors but rather the appearance of new, extrajunctional ACh receptors.

ring in the muscle tissue at the height of denervation supersensitivity is about 20 times the number of synaptic receptors or about 5×10^8 per muscle fiber and corresponds to about 600 receptor sites per μm^2 of surface membrane. Because both the receptor density and the acetylcholine sensitivity have been measured in rat diaphragm at various times after denervation, it has been possible to correlate these parameters: the sensitivity is roughly proportional to the square of the receptor density (Hartzell and Fambrough, 1972). This proportionality will be used later in this chapter to transform physiological data into quantitative molecular terms for analysis of the down-regulation of receptors by muscle activity.

When the first maps of the distribution of chemosensitivity along denervated muscle fibers were constructed, it was apparent that the large increase in extrajunctional acetylcholine sensitivity occurred without loss of junctional sensitivity. Nevertheless, the term 'spread of sensitivity' was used to describe the change following denervation, and this term suggests to some that the underlying mechanism for the change is a redistribution of junctional receptors into the extrajunctional regions. There is abundant evidence that this does not happen to any appreciable extent. On the contrary, it is now clear that denervation supersensitivity involves the biosynthesis and insertion of newly synthesized ACh receptors into extrajunctional plasma membrane.

There have been several direct demonstrations that the ACh receptors appearing after denervation are newly synthesized. Hall and co-workers have labeled denervated rat diaphragms in organ cultures with [^{35}S]methionine, isolated cholinergic receptors by lectin binding and affinity chromatography, and shown that during periods when extrajunctional sensitivity was increasing, the receptor fraction contained very much more radioisotopic label than the control fraction (Brockes and Hall, 1975). Devreotes and Fambrough (1976) have demonstrated the incorporation of heavy isotopically labeled amino acids into receptors of denervated mouse extensor digitorum longus (EDL) muscle. Linden and Fambrough (1979) have shown the same for denervated rat EDL and soleus muscles.

The experiments involving heavy isotope labeling are worth describing in further detail here both because the technique has been applied to the study of ACh receptors down-regulation and because the technique is applicable to various receptor systems and is currently being used to study down-regulation of insulin receptors and regulation of β-adrenergic receptors. Certain algal strains have been adapted to grow in deuterated water and utilize oxides of carbon 13 and nitrogen 15 as substrates for biosynthesis of all their organic molecules. Acid hydrolysates of the resulting proteins contain a mixture of most of the amino acids used for protein synthesis by vertebrates. Tissue or organ culture medium containing such amino acids (in place of the ordinary ^{1}H, ^{12}C, ^{14}N-containing amino acids) can be used for *in vitro* maintenance of denervated muscle. The muscle will utilize the exogenous, isotopically labeled amino acids for construction of new protein molecules. The heavy isotope-labeled proteins differ from normal proteins only in the presence of an extra neutron in many of the nuclei of the carbon, nitrogen and hydrogen atoms. ACh receptors with close to 80 per cent substitutions of these isotopes for the more commonly occurring ones function normally (Devreotes and Fambrough, 1976). While the shapes and sizes of the heavy-labeled proteins are unaltered by the additional neutrons, the weight of labeled proteins is increased. Thus the density of the molecules is greater. The 'density-shifted' molecules can be physically separated from the naturally occurring forms, when techniques are employed with fractionate on the basis of density. Two such techniques which are routine for biochemists are velocity sedimentation and buoyant-density centrifugation. In both cases the larger the molecules being separated the faster the separation and the higher the resolution. Fortunately, many different kinds of receptor molecules seem to be large proteins or glycoproteins. Included in this category are the receptors for steroid hormones, epidermal growth factor receptors, insulin receptors and cholinergic receptors.

For her studies of ACh receptor biosynthesis in denervated rat EDL and soleus muscles, Linden transferred denervated muscles into organ culture. The muscles were stretched to normal resting length and connected via one tendon to strain gauges. At appropriate times the health of the cultured muscle was determined by measuring the parameters of the muscle twitch and tetanus in response to electrical stimulation. When grown in a standard culture medium at $37°$ C in a high oxygen atmosphere the stretched muscles retained their strength for several days and developed supersensitivity to acetylcholine, producing large numbers of extrajunctional ACh receptors. When the medium contained the dense amino acids, the muscle fibers produced 'heavy' ACh receptors. This was demonstrated by blocking pre-existing ACh receptors with α-bungarotoxin, changing the medium to one containing the heavy amino acids and continuing organ culture for a period of hours, during which the muscle fibers produced new ACh receptors. The newly appearing ACh receptors were converted into receptor–bungarotoxin complexes, using ^{125}I-labeled bungarotoxin this time. The toxin–receptor complexes were solubilized from the muscle tissue and subjected to velocity sedimentation in sucrose gradients, separating the normally occurring ACh receptors from the denser heavy-isotope labeled ones. By curve-fitting the two ^{125}I-toxin–receptor peaks, the amount of newly synthesized heavy receptor was determined and the biosynthetic rate approximated as the amount of heavy receptor divided by the number of hours of incorporation of heavy receptors into plasma membrane. It is clear, even from a brief study of the curves in Fig. 6.4, that most of the receptor sites appearing after the blockade with unlabeled bungarotoxin are the heavy type, indicating that the block was effective and the biosynthetic generation of new receptor sites was occurring at an easily measurable rate in the organ cultured, denervated muscle. The size of the difference in sedimentation velocity between light and heavy receptors can be used to determine the contribution of the heavy amino acids to the heavy receptors. Analysis by Devreotes *et al.* (1977) led to the conclusion that this difference in sedimentation velocity required the heavy amino acids be used to 80 per cent of the theoretical maximum extent in biosynthesis of receptor protein. (Presumably the muscle derived the remaining 20 per cent of its amino acids for protein synthesis from breakdown of serum proteins and turnover of its own proteins plus *de novo* amino acid synthesis.)

Summarizing, the appearance of denervation supersensitivity to acetylcholine in skeletal muscle has been shown to involve a large increase in total number of receptor sites, measured by bungarotoxin binding. The new recep-

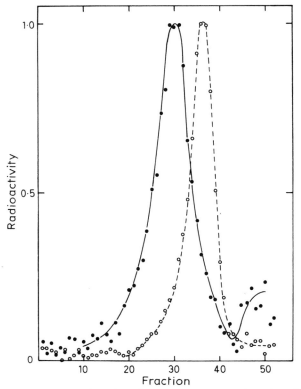

Fig. 6.4 A demonstration that the receptors appearing after denervation are newly synthesized. Denervated rat EDL muscles were cultured in normal organ culture medium or in medium containing heavy amino acids instead of those of usual isotopic composition. The receptors appearing in the plasma membrane during the culture period were labeled with different radioactive derivatives of α-bungarotoxin, solubilized and the solutions mixed and subjected to velocity sedimentation in a sucrose gradient. Sedimentation was from right to left. The 'heavy' receptor–toxin complexes (●) and normal receptor–toxin complexes (○) are well-resolved. Radioactivity in the two peaks was normalized to give the same peak height. (From Linden and Fambrough, 1979.)

tors have been shown by both electrophysiological and radioautographic methods to be distributed fairly evenly over the extrajunctional surface of the muscle fibers. Biophysical and biochemical techniques have been used to demonstrate the incorporation of isotopically labeled amino acids into receptor protein at a rate sufficient to account for the production of the extra receptors which accumulate in denervated muscle. Since the development of denervation supersensitivity involves new receptor biosynthesis, one might expect that inhibitors of protein synthesis would block this development. Indeed, it was shown some years ago that inhibitors of either RNA or protein synthesis can block the development of denervation supersensitivity (Fambrough, 1970).

The regulation of ACh receptor packing density in the extrajunctional plasma membrane of denervated skeletal muscle is not simply a matter of a fixed number of receptors appearing in response to denervation. The median 'life-span' for the extrajunctional ACh receptor molecules is less than 24 hours. There is continuous biosynthesis and degradation of receptor molecules, the population density of receptors in the plasma membrane being set by the balance of these opposing processes.

The most detailed understanding of the metabolism of receptor molecules has come from analysis of metabolic processes in tissue-cultured skeletal muscle derived from embryos. Embryonic muscle fibers resemble denervated muscle fibers in that they have a large population of extrajunctional receptors. Muscle fibers which form *in vitro* from tissue-cultured myoblasts also have a large population of extrajunctional receptors. The tissue-cultured material is especially amenable to the study of receptor metabolism because the investigator is able to manipulate the culture conditions and because hundreds of identical cultures can be employed in a single experiment, minimizing many of the errors of measurement. Much of the information about receptor metabolism is beyond the scope of this chapter and has recently been summarized. We will only briefly state the essential features of the underlying processes. This much is needed as background for later discussion of receptor regulation.

A schematic representation of the 'life history' of receptor molecules is presented in Fig. 6.5. Receptor molecules (glycoproteins with molecular weight about 250 000 daltons) are composed of sets of polypeptide chains, probably four different kinds of chains per receptor unit. These are synthesized and rapidly assembled into receptors, which are membrane-associated (Devreotes and Fambrough, 1975; Devreotes *et al.*, 1977). Newly synthesized receptor molecules resemble the functional receptors of the plasma membrane in molecular weight, ligand binding specificity and other major physicochemical properties. However, the newly synthesized molecules are located mostly in the Golgi apparatus (Fambrough and Devreotes, 1978). The initial assembly and appearance of the receptor sites in the Golgi requires less than 30 minutes and possibly less than 15 minutes. The new receptor molecules reside in the Golgi apparatus for about two hours and subsequently are transported to the plasma membrane. The total intracellular residence time is about three hours, and about 12 per cent of the total cell receptors are in this intracellular pool of newly synthesized molecules. Insertion into the plasma membrane is an energy-dependent process and probably involves the

HYPOTHETICAL 'LIFE-CYCLE' of ACETYLCHOLINE
RECEPTORS

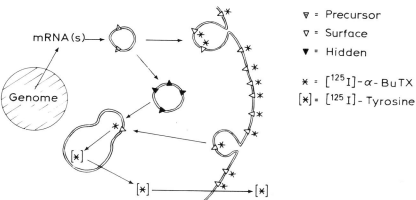

Fig. 6.5 Schematic life history of extrajunctional ACh receptors. The figure depicts a cross-section through a muscle fiber with ACh receptors symbolized as triangles in membrane profiles. The newly synthesized receptors ($\triangledown$) occur in the Golgi apparatus and post-Golgi vesicles. Hidden receptors ($\blacktriangledown$) occur on internal membranes and apparently do not cycle in and out of the plasma membrane. ACh receptors in the plasma membrane ($\triangledown$) occur as individual molecule units and also as clusters. These receptors are depicted as associated with iodinated α-bungarotoxin ($\star$) as would occur in an experiment to estimate turnover rate or count number of surface receptors. The degradation pathway is shown, transporting toxin–receptor complexes to lysosomes, where proteolysis destroys the toxin and receptor, liberating iodinated tyrosine, , which is free to diffuse out of the muscle. Although the cartoon implies fusion of new membrane with plasma membrane, budding of membrane vesicles off plasma membrane for transport to lysosomes and fusion of such membrane with lysosomes, these aspects are conjectural; they have not been demonstrated directly.

fusion of post-Golgi vesicles with the plasma membrane, exposing the binding sites of the receptor molecules to the extracellular space. About 75 per cent of the cell-associated receptors occur in this exposed form. The surface receptors have a median lifetime of about 20 hours and are destroyed by a mechanism which involves interiorization of receptor-containing membrane, transport to the lysosomal compartment and proteolytic destruction (Devreotes and Fambrough, 1975; Fambrough *et al.*, 1977). The remaining 13 per cent of receptors in the system are 'hidden' receptors which metabolically are not distinguishable from the surface receptors. However, this population is intracellular. There is strong evidence that receptors in this population are not cycling into and out of the plasma membrane (Gardner and Fambrough, 1979). The significance of the 'hidden' receptor population is unknown, and its existence *in vivo* is not well established.

There are many lines of experimental evidence which have led to this description of receptor metabolism. The most satisfying data have come from direct labeling studies in which heavy-isotope labeled amino acids were used to label receptor molecules directly. These studies are illustrated here in two figures. Fig. 6.6 shows the kinetics of appearance of newly synthesized receptor molecules in the surface membrane. While there is continuous incorporation of receptor molecules into the plasma membrane (open circles), these molecules do not contain the labeled amino acids (given to the cells from time zero) until several hours have passed. During this time the intracellular population of receptors in the biosynthetic pathway become fully labeled. Then, about three hours after feeding the cells the amino acids of heavy isotope composition, receptors rich in aminoacyl residues containing the heavy isotopes begin to appear in the plasma membrane (closed circles). After a slightly longer time, all of the receptors appearing in the membrane are of the

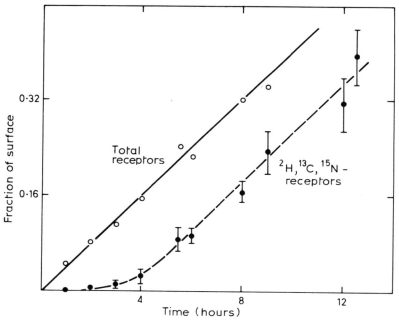

Fig. 6.6 Kinetics of appearance of total (^{1}H-, ^{12}C-, ^{14}N- plus ^{2}H, ^{13}C, ^{15}N-) and density shifted (^{2}H, ^{13}C, ^{15}N-) receptors in the plasma membrane during incubation from time zero in medium containing ^{2}H, ^{13}C, ^{15}N-amino acids. (From Devreotes *et al.*, 1977.)

isotopically labeled variety. This is signified in Fig.6.6 when the slopes of the two curves become identical. When the kinetics of appearance of new receptor sites in the plasma membrane is examined over a moderate time-period, there appears to be a *linear* addition of new molecules to the plasma membrane with time. This is consonant with the known mechanisms of protein synthesis, wherein a certain number of messenger RNAs are translated at a steady rate to yield a certain number of polypeptides per unit time. We will use the sumbol r to characterize the rate of receptor production and appearance in the plasma membrane. This constant r is the biosynthetic rate and has

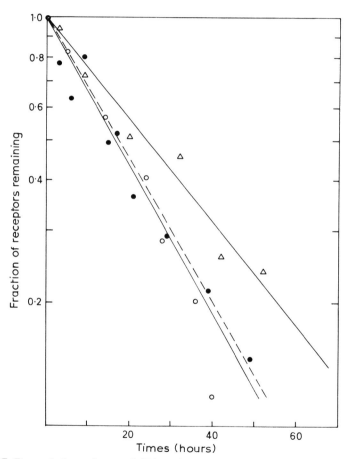

Fig. 6.7 Degradation of normal (●) and density-shifted (○) ACh receptors without bound α-bungarotoxin, employing a pulse–chase experimental regime, and a parallel time course of degradation of [^{125}I]-α-bungarotoxin–receptor complexes (△). (Data from Gardner and Fambrough, 1979.)

units *receptors per hour* (when considering a fixed number of muscle fibers or a unit area of plasma membrane and surface area changes are inconsequential).

The kinetics of receptor degradation are different from the kinetics of production (Fig. 6.7). Degradation has the kinetics of a random-hit process. It is as though the muscle fiber internalizes and hydrolyses about 3 per cent of its surface membrane each hour, for about 3 per cent of the receptor sites in the plasma membrane are destroyed by interiorization and hydrolysis every hour. There is no preferential destruction of newer or older molecules. That is, a newly synthesized receptor molecule newly arrived at the cell surface stands the same chance of being degraded in the next unit of time as does the ancient receptor molecule which has, by chance, survived in the plasma membrane for several times the lifetime of the average receptor molecule. Clearly the turnover of receptors has nothing directly to do with receptor function. Receptors are degraded at the same rate whether or not their binding sites for ligands are occupied by agonist, antagonist or left unoccupied (Gardner and Fambrough, 1979).

We can formalize the kinetics of degradation of receptors as a simple first-order decay process with rate constant k (units are reciprocal time). If the initial receptor population consists of R_0 receptors, then after time t the number of these same receptor molecules which have survived (R_t) is given by

$$R_t = R_0 e^{-kt} \qquad (6.1)$$

The development of denervation supersensitivity is a reversible process. The severed nerve has a propensity for regrowing to the target tissue and re-establishing connections with it. Soon after functional connections are re-established in skeletal muscle, the extrajunctional sensitivity declines to normal levels. This decline has been interpreted as a consequence of the resumption of neuronally driven muscle activity or as a consequence of some humoral signal from the reinnervating nerve, a signal independent of muscle activity. Probably both mechanisms exist, the importance of each in the overall decline of sensitivity varying from one muscle to another and from one species to another. This matter has been the subject of many reviews, including a recent, lengthy one by this author (Fambrough, 1979). Resolution of the matter remains a goal of many research programs. For the present discussion, we shall bypass the issue of what stimuli trigger the decline in extrajunctional ACh sensitivity and turn to the question of the mechanisms mediating the decline.

Some observations by Lømo and Rosenthal (1972) convinced the scientific world that a nearly complete elimination of extrajunctional sensitivity of denervated muscle could be experimentally induced by chronic electrical stimulation of the muscle, even in the absence of reinnervation. These observations have been repeated and extended in many laboratories, including important work by Lømo and Westgaard (1975) which describes the parameters of electrical stimulation which are effective in reducing extrajunctional sensitivity and the kinetics of this reduction.

Lømo and Westgaard (1975) used chronically implanted electrodes to stimulate denervated soleus and extensor digitorum longus (EDL) muscles of unrestrained rats. The muscles had been denervated for a period sufficient to allow full development of denervation supersensitivity before the electrical stimulation was begun. Acetylcholine sensitivity of the muscle fibers was measured by conventional electrophysiological techniques. The most effective regime of stimuli was a pattern of fairly high frequency shocks for about a second given at intervals of a few minutes. With this regime acetylcholine sensitivity fell from several hundred millivolts per nanocoulomb to less than 50 mV/nC in two days. Converting the sensitivity values into relative numbers of acetylcholine receptors per unit area of membrane, this corresponds to the loss of approximately two-thirds of the receptors. Such a rapid change in the number of receptors could be accomplished by acceleration of the degradation rate or by inhibition of receptor biosynthesis or by an appropriate combination of these two effects. If the rate of receptor degradation in the absence of electrical stimulation results in destruction of half of the starting population of receptor molecules after 24 hours (a reasonable estimate), then we can state the above more quantitatively. The loss of two-thirds of the receptors in two days would require acceleration of the degradation rate by twenty-fold without change in biosynthetic rate. Alternatively, nearly complete inhibition of receptor biosynthesis without appreciable change in degradation rate would accomplish this same change in receptor number.

A fair approximation of the rate of ACh receptor degradation can be obtained be measuring the rate of proteolytic destruction of ^{125}I-labeled α-bungarotoxin which has bound to the receptor sites (Devreotes and Fambrough, 1975; Gardner and Fambrough, 1979). This method has been used to investigate the effect of electrical stimulation upon the turnover of extrajunctional acetylcholine receptors in denervated rat diaphragm and leg muscles (Hogan *et al.*, 1976; Linden and Fambrough, 1979). In no case has an acceleration of degradation rate been observed. The effect of electrical stimulation upon the biosynthesis of receptors in these same muscles has been studied by measuring the incorporation of isotopically labeled amino acids into receptor molecules (Reiness and Hall, 1977; Linden and Fambrough, 1979). Electrical stimulation, given in patterns which were documented by Lømo and Westgaard (1975) to be most effective in reducing extrajunctional chemosensitivity, greatly reduced the incorporation of labeled amino acids into receptors. Linden and Fambrough (1979) examined the kinetics of onset of this inhibition of receptor biosynthesis and performed control experiments which demonstrated that the stimulated muscles carried out protein synthesis at a rate comparable to unstimulated muscles and maintained their strength of contraction during the experiments. The results showed a selective inhibition of receptor biosynthesis and incorporation into plasma membrane. The inhibition had a fairly abrupt onset at about 18 hours after the beginning of electrical stimulation (Fig. 6.8). In semi-quantitative terms, all of these experiments led to the conclusion that the loss of extrajunctional ACh sensitivity in denervated muscles activated by chronic electrical stimulation is mediated by a selective inhibition of receptor biosynthesis without a major effect upon the degradation of receptors. One can infer from this that the absence of extrajunctional receptors in normal adult, innervated muscle results from a continuous repression of receptor biosynthesis in the extrajunctional portions of the muscle fibers.

Before leaving this point about inhibition of extrajunctional receptor

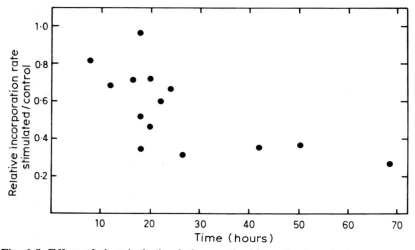

Fig. 6.8 Effect of electrical stimulation on the biosynthesis and plasma membrane of incorporation of ^2H, ^{13}C, ^{15}N-labeled ACh receptors in denervated rat EDL muscle in organ culture. The muscles contracted vigorously with full tetanic tension in response to electrical stimulation (100 Hz for 1 s given every 80 s). Relative biosynthetic rate was measured as the amount of heavy receptors (identified by velocity sedimentation as illustrated in Fig. 6.4) synthesized per unit time per mg of muscle weight in stimulated muscles divided by that measured in non-stimulated muscles cultured for the same time. Cultures were labeled with ^2H, ^{13}C, ^{15}N-labeled amino acids for 7–9 hours in experiments where stimulation was for 8–18 h. All longer experiments had labeling times of 11–12 hours. Each point on the graph represents one experiment, involving two or three stimulated muscles and two or three control, unstimulated muscles. (From Linden and Fambrough, 1979.)

biosynthesis, let us consider some of the above-mentioned results in a more formal way. If a cell lacked receptors in its plasma membrane and suddenly began to produce them at a rate of r receptors per hour, eventually the number of receptors in the membrane would plateau at a level determined by r and by the degradation rate. In fact, given the kinetics of degradation described by Equation (6.1), the number of receptors when a steady state is reached would be equal to r/k and the kinetics of approach to the steady state are described by the equation:

$$R_t = \frac{r}{k} (1 - e^{-kt}). \tag{6.2}$$

Now, consider a cell which already has R_0 receptors when we first become interested in it and which from our first moment of interest has a biosynthetic rate r and degradation rate constant k. New receptors will populate the plasma membrane with kinetics given by Equation (6.2) while the original population R_0 will disappear with kinetics given by Equation (6.1). Thus, at any time t the number of receptors R_t will be the sum of these:

$$R_t = \frac{r}{k} (1 - e^{-kt}) + R_0 e^{-kt} \tag{6.3}$$

The same equation holds no matter whether at time zero the biosynthesis rate shifts from zero to some finite rate (as in Equation 6.2), remains unchanged, or drops to a lower value. Thinking, now, about the effect of electrical stimulation on extrajunctional receptors of denervated muscle, we imagine that r decreases while k remains unchanged as a result of muscle activity driven by the stimuli. That is a conclusion of Linden's experiments. The biosynthetic rate r dropped after about 18 hours of stimulation, and the drop was abrupt. The prediction, then, would be that the number of extrajunctional receptors would remain unchanged during the first 18 hours of stimulation and then would decrease exponentially toward some new, lower number. The characteristic time for the decrease would be $1/k$ and the final receptor number r/k.

There is some evidence that this is, indeed, the kinetic path for loss of extrajunctional ACh sensitivity. When the data of Lømo and Westgaard (1975) are translated into molecular-numerical terms and the points fit to Equation (6.3), assuming a value of k typical to those measured by the degradation of bound toxin method mentioned above, it is apparent that the exponential decline in receptors should begin hours to days after the onset of electrical stimulation, the time of onset depending upon the pattern of stimulation (Fig. 6.9). To my knowledge, no tests have been made to confirm or substantiate the validity of these back-extrapolations (except for the measurements of r made by Linden). Perhaps the most important inference one can draw from these considerations is that the effect of muscle stimulation on receptor biosynthesis may operate by a cumulative influence which eventually

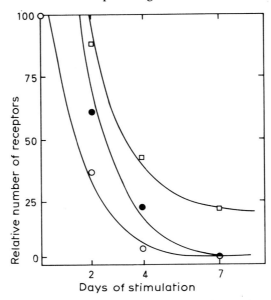

Fig. 6.9 Decline in number of ACh receptors in denervated rat soleus muscles due to electrical stimulation. Data on ACh sensitivities (Lømo and Westgaard, 1975) were converted into relative number of ACh receptors, using the direct proportionality between ACh sensitivity and the square of the packing density of ACh receptors (Hartzell and Fambrough, 1972; Fambrough, 1974). Symbols refer to three patterns of electrical stimulation: 1 Hz continuous (□); 10 Hz for 1 s given once every 10 s (●); and 100 Hz for 1 s given once every 100 s (○). The curves, fitted by eye to the experimental points, are all first-order exponentials with a half-time of 24 hours (a generous estimate of the typical lifespan of an extrajunctional ACh receptor). These are the curves predicted for the decline in receptors if such a decline were due to complete or partial inhibition of biosynthesis of receptors with no significant change in degradation rate. In order to fit the data, the time at which the biosynthetic rate changes in response to electrical stimulation must lag many hours behind the onset of stimulation. (From Fambrough *et al.*, 1977.)

may exceed some threshold and trigger an abrupt inhibition of receptor production. From what we know about the biochemistry of muscle contraction and the permeability changes accompanying muscle action potentials, one can imagine that chronic electrical stimulation would lead to changes in levels of many compounds, including nucleosides and nucleotides, cyclic nucleotides, creatine, and also the concentrations of most inorganic ions. The mechanism for repressing receptor biosynthesis might be coupled to any one of these changes. Given the difficulty in establishing linkage between the immediate effects of muscle activity and the delayed (and poorly understood) repression of receptor production, I expect the intermediates in this repression will be discovered only serendipitously.

The fact that the nervous influence on muscle ACh receptors involves elimination of receptors distant from the nerve terminals and lasting establishment of receptor-rich areas nearest to the nerve terminals has been called a paradox and has led to various speculations about the underlying mechanism. Among the more interesting ideas have been that (1) the synaptic ACh receptors might be synthetic products of the nerve, transferred to the muscle, (2) the synaptic receptors might be products of a different set of genes than the genes coding for extrajunctional ACh receptors and these synaptic receptor genes might be under a different control mechanism, and (3) the establishment of positional and metabolic stability might be important in the differential accumulation of receptors in the junctional regions. At present there is little evidence to support any of these ideas. In fact, there are some reasonable arguments that favor biosynthesis of both junctional and extrajunctional receptors in the muscle fibers rather than in neurons and there is direct evidence that extrajunctional ACh receptors can become part of the post-synaptic organization at newly forming junctions. Space does not permit discussion in this Chapter of these matters or of other aspects of ACh receptor regulation such as the turn-on of ACh receptor biosynthesis during myogenesis, the clustering of receptor sites in non-innervated muscle fibers, the 'down-regulation' of ACh receptors by carbachol and anti-cholinesterase drugs, and the effects of anti-ACh receptor antibodies and other anti-receptor immune responses on receptor turnover and function and the relation of these processes to myasthenia gravis. Most of these matters have been reviewed recently (see Fambrough, 1979, and references therein). What has become very clear is that many of the processes currently capturing the curiosity of cell biologists are processes which can involve ACh receptors.

In the broader context of receptors, the acetylcholine receptors of skeletal muscle have yielded most, so far, to the assaults by scientists. In fact, I believe the receptor concept itself can be attributed to Langley (1905) and his studies of nicotinic ACh receptors. The experimental strategies which have succeeded with ACh receptors can, in many cases, be applied with only slight modifications to other receptors systems. For example, the down-regulation of insulin receptors, EGF receptors, or steroid receptors could be examined from a metabolic point of view through experiments employing dense-isotope labeling.

Simple kinetic modeling of receptors can be applied to regulatory processes governing receptor number to yield some satisfactory semi-quantitative accounting of receptor number as a function of time. It may be surprising that as simple a kinetic formalism as Equation (6.3) is adequate for fitting much of the data on ACh receptor regulation. In Equation (6.3), the values of biosynthetic rate r and degradation rate constant k were fixed at time zero. A simple elaboration of Equation (6.3) is to make r and k continuous functions of time. A further addition, useful in some systems, is to divide the set of receptors which exist at any particular time (R_t) into two pools (which might represent, for example, the plasma membrane-associated and internal receptors of a cycling system or the cytosolic and nuclear receptors in steroid hormone shuttling). For a cycling system, then, $R_t = R_{exterior} + R_{interior}$ and the balance between exposed and interiorized receptors might be modeled with interiorizing and exteriorizing rate constants. The importance of modeling lies partly in identifying what parameters of receptor regulation may be necessary and sufficient to account for the observations. Such models also aid in formulating further experiments as, for example, analysis of the time-course of repression of ACh receptor production which results from muscle acitvity.

REFERENCES

Albuquerque, E.X. and McIsaac, R.J. (1970), *Exp. Neurol.*, **26**, 183–202.

Axelsson, J. and Thesleff, S. (1959), *J. Physiol. (London)*, **147**, 178–193.

Brockes, J.P. and Hall, Z.W. (1975), *Proc. Natl. Acad. Sci. USA*, **72**, 1368–1372.

Cannon, W.B. and Rosenblueth, A. (1949), *The Supersensitivity of Denervated Structures*, Macmillan, New York.

Devreotes, P.N. and Fambrough, D.M. (1975), *J. Cell Biol.*, **65**, 335–358.

Devreotes, P.N. and Fambrough, D.M. (1976), *Proc. Natl. Acad. Sci. USA*, **73**, 161–164.

Devreotes, P.N., Gardner, J.M. and Fambrough, D.M. (1977), *Cell*, **10**, 365–373.

Fambrough, D.M. (1970), *Science*, **168**, 372–373.

Fambrough, D.M. (1974), *J. Gen. Physiol.*, **64**, 468–472.

Fambrough, D.M. (1979), *Physiol. Rev.*, **59**, 165–227.

Fambrough, D.M. and Devreotes, P.N. (1978), *J. Cell Biol.*, **76**, 237–244.

Fambrough, D.M., Devreotes, P.N. and Card, D.J. (1977), in *Synapses* (G.A. Cottrell and P.N.R. Usherwood, eds), Blackie, Glasgow.

Fertuck, H.C. and Salpeter, M.M. (1974), *Proc. Natl. Acad. Sci. USA*, **71**, 1376–1378.

Gardner, J.M. and Fambrough, D.M. (1979), *Cell*, **16**, 661–674.

Hartzell, H.C. and Fambrough, D.M. (1972), *J. Gen. Physiol.*, **60**, 248–262.

Hogan, P.G., Marshall, J.M. and Hall, Z.W. (1976), *Nature*, **261**, 328–330.

Langley, J.N. (1905), *J. Physiol. (London)*, **33**, 374–413.

Lee, C.Y. (1972), *Annu. Rev. Pharmacol.*, **12**, 265–286.

Linden, D.C. and Fambrough, D.M. (1979), *Neuroscience*, **4**, 527–538.

Lømo, T. and Rosenthal, J. (1972), *J. Physiol. (London)*, **221**, 493–513.

Lømo, T. and Westgaard, R.J. (1975), *J. Physiol. (London)*, **252**, 603–626.

Miledi, R. (1960), *J. Physiol. (London)*, **151**, 1–23.

Neher, E. and Sakmann, B. (1976), *J. Physiol. (London)*, **258**, 705–729.

Pecot-Dechavassine, M. (1968), *Arch. Int. Pharmacodyn.*, **176**, 118–133.

Reiness, C.G. and Hall, Z.W. (1977), *Nature*, **268**, 655–657.

7 Insulin Receptors in Disorders of Glucose Tolerance and Insulin Sensitivity

EMMANUEL VAN OBBERGHEN, FLORA DE PABLO and JESSE ROTH

Receptor Regulation
(*Receptors and Recognition*, Series B, Volume 13)
Edited by R. J. Lefkowitz
Published in 1981 by Chapman and Hall, 11 New Fetter Lane, London EC4P 4EE
© 1981 Chapman and Hall

7.1 INTRODUCTION

It has long been known that glucose stimulates the β-cells of the pancreas to release insulin which acts on target cells to promote glucose utilization and storage. Moreover, only glucose could be measured so that the system was oversimplified – hyperglycemia was due to insulin deficiency and hypoglycemia was due to insulin excess. With the advent of insulin radioimmunoassay it became clear that only a minority of patients conform to this scheme (Yalow and Berson, 1960). The diabetics who are truly insulin-requiring are indeed insulin-deficient, but the majority of hyperglycemic patients are not absolutely insulin-dependent and have normal or supernormal concentrations of circulating insulin. Since the plasma insulin in these subjects was shown not to be bound to any macromolecules (e.g. to insulin antibodies or to other binding proteins) and was qualitatively normal, these patients had insulin resistance at the tissue level.

Conversely, in patients with hypoglycemia due to islet cell tumors, insulin levels were found to be inappropriately elevated, which conformed to expectations, while in many patients (e.g. growth hormone deficiency; anorexia nervosa) with low normal glucose or frank hypoglycemia, plasma insulin was noted to be low or unmeasurable. The introduction of methods to measure the receptor has resolved many of these apparent paradoxes. Indeed, it was shown in many disorders of glucose metabolism, where hormone levels were discordant with the clinical state, that the receptor was altered and reflected the clinical state. These findings enable us now to divide the disorders of glucose metabolism in humans into two groups (Table 7.1).

In the first group the clinical picture correlates well with the target cell characteristics, but the hormone levels are discordant with the biological observations. In the second group of clinical states, the hormone dominates and is concordant with the insulin levels, while the receptor only modifies the clinical state.

7.2 DISORDERS WITH DISCORDANT HORMONAL LEVELS

7.2.1 Insulin resistant states

(a) Hyperglycemia with moderate insulin resistance

(i) Diabetes mellitus
Studies of insulin receptor have been performed in both thin and obese adult-onset (insulin-independent) diabetics. The majority of adult-onset

Table 7.1 Concordance and discordance or circulating hormone concentrations and the biological state

	Hormone concentration subnormal or unmeasurably low	Hormone concentration normal or elevated
	Concordant	Discordant
Biological response deficient	Insulin-dependent (juvenile type) diabetes (~20%)	Non-insulin dependent (adult type) diabetics (~80%)
		Growth hormone excess
		Glucocorticoid excess
		Uremia
		Syndromes of extreme insulin resistance
	Discordant	Concordant
Biological response excessive	Nonislet cell tumors Anorexia nervosa	Islet cell tumors Infants of diabetic mothers
	Growth hormone deficiency	
	Glucocorticoid deficiency	

diabetics are obese and frequently insulin resistant. Regardless of the severity of the hyperglycemia, these obese patients demonstrate the same inverse relationship between receptor concentration and basal levels of insulin (see later). Olefsky and coworkers studied a group of thin, insulin-dependent diabetics, and found the same inverse relationship between receptor concentration and ambient levels of insulin (Olefsky and Reaven, 1974, 1977). In this group of thin diabetics Olefsky and Reaven (1977) also noted an improvement in insulin binding after treatment with an oral sulfonylurea. Because the improved binding occurred without significant change in insulin levels, it was postulated that the sulfonylurea may act by affecting directly the insulin–receptor interaction. However, in a few patients sulfonylureas were without effect on the insulin receptors, and the same patients failed to respond clincially to the drug. The effect of sulfonylureas in increasing insulin receptor concentrations has also been demonstrated in rodents. Thus, contrary to long-held theories, sulfonylureas, when they work, appear to operate at the level of the receptor rather than at the level of the hormone. The long-ignored paradox had been that sulfonylureas do act acutely to stimulate hormone release *in vivo* and *in vitro*, but, curiously, during chronic treatment of diabetics with sulfonylureas, circulating hormone concentrations failed to increase. Nevertheless, even if sulfonylureas do appear to act to increase receptors in receptor-deficient patients, we do not recommend the use of these agents in the treatment of diabetes; the mechanism of their action is a pharmacological question, while their clinical usefulness is a therapeutic question. The answer to this question depends on their safety (i.e. side-effects) and long-term efficacy, which have been in some doubt.

(ii) Acromegaly

A regulatory role of insulin on its own receptor concentration has been suggested in acromegaly, a clinical state characterized by elevated levels of circulating growth hormone as well as hyperinsulinemia and insulin resistance. We have found a direct relationship between plasma growth hormone and plasma insulin and an inverse relationship between ambient levels of basal (fasting) insulin and the concentrations of insulin receptors on circulating monocytes. However, there were also associated striking changes in receptor affinity, which may play a key role in determining whether glucose tolerance is severely distributed (Muggeo *et al.*, 1979a).

(iii) Obesity

Insulin receptors have been evaluated extensively in obesity, the most common hormone-resistant state in humans (Archer *et al.*, 1975; Armatruda *et al.*, 1975; Olefsky, 1976; Bar *et al.*, 1976, 1979; Harrison *et al.*, 1976). The binding of ^{125}I-labeled insulin either to monocytes or adipocytes in some obese patients is entirely normal, whereas in other patients there is a decrease in insulin binding that can be very pronounced. Scatchard analysis of the binding data indicates that the decrease in insulin binding is due entirely to a reduction of the receptor concentration (Fig. 7.1a and b); the remaining receptors are entirely normal, including their affinity for insulin. In obese humans, as in mice, there exists an excellent correlation between the magnitude of the insulin resistance and the severity of the receptor defect. Thus, when the binding data on an individual patient are compared with the clinical state of that patient, it is found that the obese patient with the lowest binding and the greatest receptor loss also suffers the most severe insulin resistance. Furthermore, obese patients who show no signs of insulin resistance have normal binding of insulin and normal insulin receptor concentrations on their monocytes and adipocytes. The inverse relationship between receptor con-

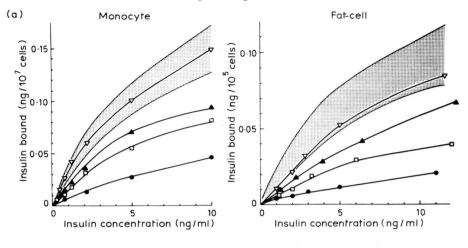

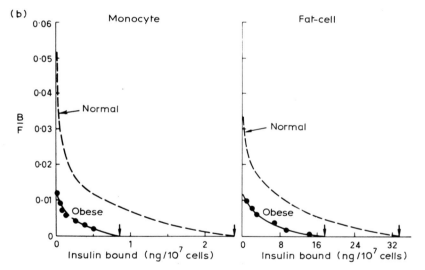

Fig. 7.1 (a) Insulin binding to circulating monocytes from four obese patients (left) and insulin binding to isolated fat-cells from four different obese patients (right). Bound insulin is plotted as a function of the amount of insulin in the incubation assay. Shaded areas are the ranges of normal (mean ± 1 s.d.). (b) Scatchard plots of insulin binding from the two obese patients in (a) with the lowest insulin binding. Dashed lines represent mean data from several studies in thin, normal subjects. Inverted arrows (abscissa intercepts) designate the maximal binding capacities and permit calculation of receptor concentrations. B/F = bound/free hormone (From R.S. Bar *et al.*, 1979). There are seven studies of insulin receptors in human obesity. Six report findings similar to those depicted here.

centration and ambient levels of insulin *in vivo* exists also during dieting (Harrison *et al.*, 1976). Thus, calorie-restricted diets lead to a prompt improvement in glucose tolerance, amelioration of the hyperinsulinemia, and restoration of receptors to normal. It should be emphasized that the improved clinical sensitivity to insulin and the increase in receptor concentrations occur before a substantial amount of weight is lost, suggesting that adiposity *per se* is not the direct cause of the insulin resistance, but that diet may be directly involved in the pathogenesis of the resistant state.

(iv) Other forms of moderate insulin resistance
Receptors are just starting to be characterized in other forms of moderate insulin resistance. Glucocorticoid excess produces insulin resistance similar to that encountered in obesity and acromegaly, and a decrease in hormone binding at the receptor has likewise been observed (Olefsky, 1975; Olefsky *et al.*, 1975; Bennett and Cuatrecasas, 1972; Kahn *et al.*, 1978). However, the effects of glucocorticoids appear to be quite complex, since acute and chronic therapy in humans and rodents, *in vivo* and *in vitro*, each appear to give somewhat different results, i.e. effects on affinity *versus* concentration of insulin receptors.

Similarly, in one form of experimental uremia, the insulin resistance is accounted for by the receptor defect, whereas chronic renal failure in humans appears to be more complicated, involving receptor and post-receptor sites. Pregnancy, myotonic dystrophy and Werner's Syndrome have likewise been studied, but the relative role of receptor and post-receptor events is as yet uncertain.

(b) Hyperglycemia with extreme insulin resistance
In contrast to the patients with moderate insulin resistance described above, there are several groups of patients with extreme insulin resistance (Table 7.2). These patients are characterized by a more-pronounced hyperin-

Table 7.2 Syndromes of extreme insulin resistance

1. Anti-receptor antibodies present
   (a) Insulin resistance and acanthosis nigricans Type B
   (b) Ataxia telangiectasia
   (c) New Zealand Obese (NZO) mouse
   (d) IgA deficiency
2. Without antireceptor antibodies

	Binding defect	Postbinding defect
(a) With acanthosis nigricans	Type A	Type C
(b) Lipoatrophic diabetes	√	√
(c) Leprechaunism	√	√

Table 7.3 Syndrome of extreme insulin resistance with acanthosis nigricans

1. Extreme elevation of endogenous plasma insulin concentrations – basal and stimulated
2. Extreme resistance to exogenous insulin
3. Absence of any previously known cause of insulin resistance, including obesity and lipoatrophy
4. Glucose tolerance ranges from normal to severe hyperglycemia with moderate ketoacidosis
5. Marked impairment of insulin binding to its receptor on freshly obtained cells (monocytes)
6. Three or more clinical subtypes:

   Type A. No immunological features

    Young females – children and adolescents

    Disordered sexual function – amenorrhea; hirsutism; enlargement of clitoris and/or labiae; polycystic ovaries

    Receptor concentration decreased; receptor affinity normal

    Accelerated early growth; no autoimmune features

   Type B. Autoimmune features

    Older; mostly females

    Immunological features – elevated erythrocyte sedimentation rate; leukopenia; alopecia; arthralgias; nephritis; antinuclear antibodies; some patients have well-defined autoimmune disease (e.g. systemic lupus)

    Antireceptor antibodies – bind to insulin receptor; block insulin binding; impair insulin action; mimic insulin action

    Clinical course – insulin resistance may be replaced by hypoglycemia; nephritis and nephrotic syndrome associated with proliferation of cellular receptors with low affinity for insulin; remissions (spontaneous or drug induced) and death in one case

   Type C. Very similar to Type A except that insulin binding to receptor is normal

sulinemia, both basal and stimulated, and a more-impaired response to exogenous insulin, when compared to the obese or other patients with moderate insulin resistance. Irrespective of the cause of the extreme insulin resistance, the skin lesion, acanthosis nigricans, is found in many of these patients (Kahn *et al.*, 1976; Table 7.3). The severity of this skin lesion fluctuates in phase with the course of the hyperinsulinemia. However, the relationship at the biochemical or cellular level of the skin lesion to the glucose and insulin problems is presently unknown.

(i) Extreme insulin resistance with anti-receptor antibodies

(ia) The syndrome of insulin resistance and acanthosis nigricans Type B. This syndrome is a rare disorder, but has provided many useful insights into the

role of receptors in disease, as well as basic information about the receptor itself (Flier *et al.*, 1975, 1976a,b, 1977; Kahn and Harrison, 1980; Harrison *et al.*, 1980). By definition, all Type B patients have circulating antibodies that reduce the binding of insulin to its own receptors. In addition, these patients have clinical and laboratory findings that we associated with autoimmune phenomena, but only a minority of the patients fulfill the diagnostic criteria for one of the well-characterized autoimmune diseases (e.g. lupus).

The hallmark of the syndrome is extreme resistance to both endogenous and exogenous insulin. The severity of the insulin resistance correlates well with the severity of the hyperinsulinemia, the extent of the defect in insulin binding to the patient's cells and with the titer of circulating anti-receptor antibody.

To investigate the possible role of the insulin receptor in this disorder, the binding of insulin to its receptor on circulating monocytes of these patients was studied. These patients, with insulin resistance and anti-receptor antibodies, show decreased binding of insulin to its receptors and this impairment is mainly due to a decrease in receptor affinity. The existing data suggest strongly that the underlying receptor is normal (Muggeo *et al.*, 1979b,c). Indeed, by removing the anti-receptor antibodies by acid wash of the cell or plasma exchange, the decreased insulin binding returns to normal. Further, the defect is absent during remission of the disease and in cultured fibroblasts from the patients.

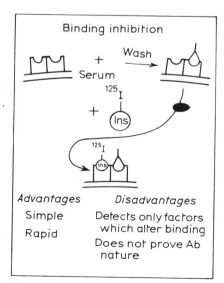 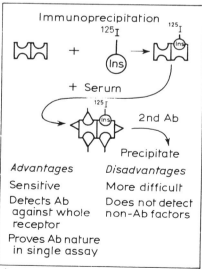

Fig. 7.2 Direct assays for insulin receptor antibodies.

To detect antibodies to the insulin receptor, two assays have been used (Fig.7.2). The binding-inhibition assay may be performed with intact cells, membranes or even solubilized receptors and has the great advantage of being very simple to carry out (Flier *et al.*, 1975). The immunoprecipitation assay has theoretically a greater sensitivity, but unfortunately is more tedious and requires solubilized receptors (Harrison *et al.*, 1979a). Until now most of

Table 7.4 Insulin-like effects of receptor antibodies

Adipocytes

Stimulation of 2-deoxyglucose transport, glucose incorporation into lipid and glycogen, and oxidation to CO_2	Kahn *et al.* (1977) Kasuga *et al.* (1978)
Stimulation of amino acid incorporation into protein	Kasuga *et al.* (1978)
Inhibition of lipolysis	Kasuga *et al.* (1978)
Activation of glycogen synthase	Lawrence *et al.* (1978)
Inhibition of phosphorylase	R. Denton (personal communication)
Activation of pyruvate dehydrogenase and acetyl CoA carboxylase	R. Denton (personal communication)
Simulation of insulin's effect on protein phosphorylation	R. Denton (personal communication)

3T3-L1 Fatty Fibroblasts

Stimulation of 2-deoxyglucose transport and glucose oxidation to CO_2	Karlsson *et al.* (1979)
Activation of lipoprotein lipase	Van Obberghen *et al.* (1979)

Muscle

Stimulation of 2-deoxyglucose transport and glucose incorporation into glycogen	LeMarchand-Brustel *et al.* (1978)
Activation of glycogen synthase	LeMarchand-Brustel *et al.* (1978)

Liver

Stimulation of amino acid (AIB) transport	A. LeCam and P. Freychet (personal communication)

Placenta

Simulation of insulin's effects on protein phosphorylation	L. C. Harrison (unpublished observation)

the antibodies against the insulin receptor have been polyclonal IgG in nature (Flier *et al.*, 1976a), although one patient has IgM autoantibodies (Harrison *et al.*, 1979b).

When studied for a variety of biological responses in different systems, the anti-receptor antibodies acutely mimicked the actions of insulin (Kahn *et al.*, 1977; Kasuga *et al.*, 1978; Lawrence *et al.*, 1978; LeMarchand-Brustel *et al.*, 1978; Karlsson *et al.*, 1979; Van Obberghen *et al.*, 1979; Table 7.4). These findings were difficult to reconcile with the insulin resistant state observed in the patients (Kahn *et al.*, 1976). However, more recent studies have shown that this insulin-like effect is only transient (Karlsson *et al.*, 1979). Thus, exposure of 3T3-L1 fatty fibroblasts to anti-receptor antibody for 36 hours results in a state of insulin resistance. This desensitization of the cells occurs without a further change in insulin binding, is specific for insulin, and appears to be an active metabolic process (Fig. 7.3).

(ib) Autoantibodies to the insulin receptor have also been detected in patients with *ataxia telangiectasia*, a genetic disorder characterized by progressive neuromuscular disease, diverse abnormalities of the immune system and insulin resistance. The autoantibodies appeared to be low-molecular-weight monoclonal IgM, and generally occur at a lower titer than in most Type B patients (Harrison *et al.*, 1979b; Bar *et al.*, 1979a).

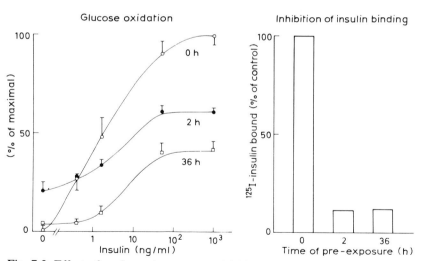

Fig. 7.3 Effect of prolonged treatment of 3T3-L1 cells with anti-receptor antibody. Cells were pretreated for 0, 2 or 36 hours with anti-receptor antibody. Insulin-stimulated glucose oxidation (left panel) and ^{125}I-insulin binding (right panel) were measured. (Adapted from Karlsson *et al.*, 1979.)

(ic) Similar antibodies have been found in the blood of the *New Zealand Obese (NZO) mouse*, a strain that has autoimmunity and insulin-resistant diabetes (Harrison and Itin, 1979).

(id) Finally, antibodies against insulin receptors are found in occasional patients with isolated *deficiency of IgA* (Harrison and Kahn, 1980).

(ii) Extreme insulin resistance without anti-receptor antibodies

(iia) *Type A*. Among the patients with severe insulin resistance and acanthosis nigricans, is a group of thin, female children or young adults, without known cause of hormonal resistance, but with problems with sexual maturation, hirsutism, and often with polycystic ovaries (Kahn *et al.*, 1976; Harrison *et al.*, 1979b; Bar *et al.*, 1980). These patients, designated as Type A, have decreased insulin binding due predominantly to a decrease in receptor concentration. By definition, these patients have no circulating antibodies or autoimmunity. Insulin binding to their monocytes tended to be much lower than previously noted in the obese patients. Neither the mechanisms responsible for these defects nor the sequence of events that leads to the receptor defect, hyperinsulinemia, and insulin resistance are known at the present time.

(iib) *Type C*. One more clinical subtype of insulin resistance associated with the skin lesions of acanthosis nigricans, Type C, should be differentiated. We have recently evaluated a young female with extreme resistance to both endogenous and exogenous insulin and with clinical features identical to those characteristic of Type A patients, including hirsutism, virilization, polycystic ovaries, and acanthosis nigricans (Bar *et al.*, 1978b). However, in contrast to the type A patients, insulin receptors on circulating monocytes of this patient were entirely normal for all tested features, i.e. her insulin receptors displayed normal concentration, affinity, specificity, and negatively cooperative interactions. In this patient, the presence of normal concentrations of insulin receptors in the setting of ambient levels of insulin, that were 30–50 times greater than normal, suggested a failure of 'down-regulation' of insulin receptors by the prevailing hyperinsulinemia. The presence of two abnormalities of insulin action, i.e. failure of insulin to lower blood glucose appropriately and failure of insulin to down-regulate its own receptor concentration, suggested the existence of a common, early post-receptor defect, perhaps involving the 'effector' system or 'second messenger'.

(iic) *Lipoatrophic diabetes*. The patients described under this rubric form a heterogeneous group of patients with glucose intolerance and severe insulin resistance. Except for the familial tendency and the lipoatrophy, which may be congenital, they resemble the Type A and C patients. Insulin binding to receptors is quite variable, being low, normal or elevated. Even within an individual family the insulin binding pattern may differ, suggesting a heterogeneity of mechanism (Wachslicht-Robard *et al.*, 1979a).

(iid) *Leprechaunism*. This syndrome, recognized in neonates, has many features, including intrauterine growth retardation, absence of subcutaneous fat, hypertrichosis, and, occasionally, acanthosis nigricans. Many, if not all leprechaun children have hyperinsulinemia and extreme resistance to insulin. In two patients the insulin binding was markedly reduced in skin fibroblasts (Schilling *et al.*, 1979) and transformed B-lymphocytes (Taylor *et al.*, 1980), suggesting major congenital or genetic defects at the receptor level. In two other patients the insulin binding to fibroblasts cultured from these patients was normal, but both appeared to have decreased insulin binding in 'freshly' isolated tissues, i.e. circulating monocytes (Kobayashi *et al.*, 1978) and liver (D'Ercole *et al.*, 1979). It seems possible that the discrepancy observed in the latter two patients between the relatively normal binding seen in cultured cells, as compared with reduced binding in fresh tissue, results from down-regulation which would occur *in vivo* in the hyperinsulinemic insulin-resistant state.

7.2.2 Hypoglycemia with supersensitivity to insulin

In five untreated patients with *anorexia nervosa* we observed abnormally low basal plasma insulin levels associated with elevated binding of insulin to the patients' red cells. After treatment, the binding of insulin and the basal plasma insulin level returned to normal values (Wachslicht-Rodbard *et al.*, 1979b).

Glucocorticoid deficiency is also associated with supernormal sensitivity to the effects of insulin. Insulin receptors in rats that are glucocorticoid-deficient have an increase in affinity without any change in receptor concentration (Kahn *et al.*, 1978). However, as we stated before, the influence of glucocorticoid appears to be quite complex and not yet well defined.

Hypophysectomized-GH-deficient patients, who have well-known supersensitivity to insulin, are found to have an increased insulin receptor concentration. In contrast, in cases of isolated GH deficiency in man, increases in affinity of insulin for receptor have been described (Lippe *et al.*, 1979).

The above examples involve instances in which the effects of one hormone – or its lack – on the receptors of another appear to have major pathological consequences. More studies are needed to unravel the variety of mechanisms by which one hormone can affect the physiology of another hormone.

7.3 DISORDERS WITH CONCORDANT HORMONAL LEVELS

7.3.1 Insulin deficiency

The situations we have discussed thus far are characterized by changes in the receptor dominating the clinical picture and by hormone levels being opposite to the clinical picture. However, even in hyperglycemic states where the dominant influence by far is the hormone deficiency, changes in receptor binding may occur and may be quite marked, even if subordinate in the ultimate to the insulin deficiency. Diabetic ketoacidosis is the most convincing example, although the mechanisms involved have not as yet been completely worked out. In this disease state insulin deficiency is the dominant problem. However, poor control (i.e. hyperglycemia), ketone bodies, and acidosis all exert major effects on the receptor. It is well established that a drop in pH from 7.4 to 6.8, which can be seen in ketoacidosis, leads to a considerable reduction in the affinity of the receptor for insulin. Potentially more important observations are that moderate acidosis itself is able to decrease the levels of receptors, and, that severe acidosis (pH <7.0) can lead to the complete disappearance of insulin receptors from the cell surface. This mechanism might explain the fact that occasionally diabetics with severe ketoacidosis fail to respond to insulin until fluid and electrolyte corrections have been initiated. Our interpretation is as follows: insulin deficiency causes severe acidosis, which leads to the disappearance of receptors and failure to respond to insulin *via* a receptor mechanism.

7.3.2 Insulin excess

We will briefly analyse two situations: (1) the patients with islet cell tumors, and (2) the infants of diabetic mothers.

(a) Insulinoma

Typically patients with insulin-secreting tumors show inappropriate increases in their basal plasma insulin levels (Bar *et al.*, 1977). Similar to obese patients, these patients have a decreased receptor concentration, which is inversely related to the plasma insulin levels. Although the cause of the insulin excess in the obese and insulinoma patients is presumably not the same, in both disease states the hyperinsulinemia appears before the receptor defect. In addition to changes in receptor concentration, some patients with insulinoma display some affinity changes. In the latter patients the speed of onset of the hypoglycemia after discontinuance of food intake seems to correlate with the receptor affinity changes.

Whether the insulin resistance found in the patients with insulinomas results from their receptor deficiency is not known at the present time. Although they are responsive to insulin as indicated by the fact that they become hypoglycemic, it is clear that they are somewhat insulin-resistant. Indeed, their plasma insulin levels are often chronically at values rarely or only transiently found in normal people. Further, these high plasma insulin levels persist for hours before leading to hypoglycemia, which usually occurs only if cessation of eating occurs simultaneously. Finally, the patients with insulinomas typically have abnormal glucose tolerance.

In conclusion, the patients with insulinomas display an inverse relationship between their receptor concentrations and their chronic circulating insulin concentration. However, at present the data do not indicate with certainty the existence of a connection between these changes and the insulin resistance.

(b) Infants of diabetic mothers
Mothers with hyperglycemia during late pregnancy often give birth to babies with a characteristic syndrome (Cornblath and Schwartz, 1976). The levels of the circulating insulin in the babies *in utero* are chronically increased, which is likely to be the consequence of an overstimulation of the infant's pancreatic β cells by unusually high levels of glucose and/or amino acids provided *via* transplacental passage from the mother. The babies are typically big, having increased weight and length. Often, they display macrosomia and an excess in the deposition of stored substrates (i.e. protein, glycogen and fat). Further, in the first few postnatal days hypoglycemia with inappropriate increases in plasma insulin is found. It is thought that the action of insulin mediated through the insulin receptor is probably responsible for the excess of stored substrates and the hypoglycemia, whereas the effects of insulin mediated through one or multiple classes of receptors with high affinity for insulin-like growth factors and low affinity for insulin probably result in macrosomia. In summary, excessive or inappropriately high levels of circulating insulin interact with the insulin receptor to produce metabolic effects, and, in addition, cross-react with receptors of closely related peptides, i.e. growth factors, to generate effects on somatic growth.

The ability of the hormone (insulin) to bind the specific receptors of other structurally related hormones (insulin-like growth factors) is referred to as 'specificity spilllover'; disease manifestations that are produced by such a crossover effect are designated as disorders of receptor design (Fig. 7.4).

An analagous situation is thought to occur in patients with non-islet cell tumors associated with hypoglycemia. In about 40 per cent of such patients, one of the insulin-like growth factors (IGF-II) is present in excesss. This insulin-like growth factor binds to its own specific receptors as well as to insulin receptors. The insulin-like growth factor is present at sufficiently high concentrations that, despite its reduced affinity for the insulin receptor, it produces activation of the metabolic effects of insulin (glucose utilization and hypoglycemia) presumably by binding to the insulin receptor. For a more complete description of this concept see Roth (1979) and Roth *et al*. (1979).

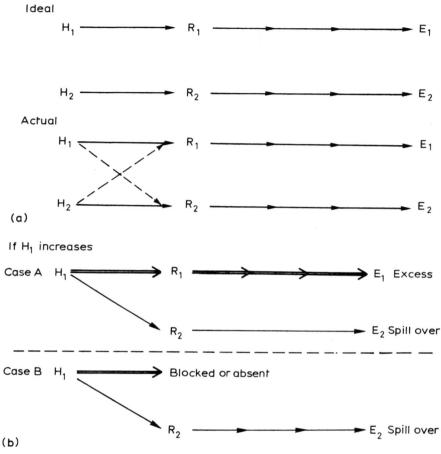

Fig. 7.4 (a) Specificity in hormone-receptor interactions. In the ideal case, each hormone (H) has its own unique receptor (R) with which it combines to produce its specific series of biological effects (E) and has no affinity at all for receptors of other hormones. Actually one hormone (H_1), in addition to its strong reactivity ($\rightarrow$) with its own receptors (R_1), can react with a weak but finite affinity ($- \rightarrow$) with receptors (R_2) of a closely related hormone (H_2); high concentrations of H_1 can thereby produce their own characteristic effects (E_1) as well as those effects (E_2) characteristic of the related hormone (H_2). (b) Specificity spillover at the receptor. In case A, the concentration of one hormone (H_1) is increased. Through its interaction with its own receptor (R_1) it produces an excess ($\Rightarrow$) of its characteristic effects (E_1) and through its interaction with the receptors (R_2) of a related hormone, it produces effects (E_2) characteristic of the other hormone (H_2). In case B the concentration of H_1 is increased but its own pathway of action is blocked or absent and the only biological effects (E_2) are those produced by the interaction of H_1 with R_2.

REFERENCES

Archer, J.A., Gorden, P. and Roth, J. (1975), *J. Clin. Invest.,* **55**, 166–174.
Armatruda, J.M., Livingston, J.N. and Lockwood, D.H. (1975), *Science,* **188**, 264–266.
Bar, R.S., Gorden, P., Roth, J., Kahn, C.R. and De Meyts, P. (1976), *J. Clin. Invest.,* **58**, 1123–1135.
Bar, R.S., Gorden, P., Roth, J. and Siebert, C.W. (1977), *J. Clin. Endocrinol. Metab.,* **44**, 1210–1212.
Bar, R.S., Harrison, L.C, Muggeo, M., Gorden, P., Kahn, C.R. and Roth, J. (1979), *Adv. Intern. Med.,* **24**, 23–52.
Bar, E.S., Levis, W.R., Rechler, M.M., Harrison, L.C., Siebert, C.W., Podskalny, J.M., Roth, J. and Muggeo, M. (1978a), *N. Engl. J. Med.,* **298**, 1164–1171.
Bar, R.S., Muggeo, M., Kahn, C.R., Gorden, P. and Roth, J. (1980), *Diabetologia,* **18**, 209–216.
Bar, E.S., Muggeo, M., Roth, J., Kahn, C.R., Havrankova, J. and Imperato-McGinley, J. (1978b), *J. Clin. Endocrinol. Metab.,* **47**, 620–625.
Bennett, V.G. and Cuatrecasas, P. (1972), *Science,* **176**, 805–806.
Cornblath, M. and Schwartz, R. (1976), in *Disorders of Carbohydrate Metabolism in Infancy,* 2nd edn, W.B. Saunders, Philadelphia, pp. 115–154.
D'Ercole, A.J., Underwood, L.E. and Groke, J. (1979), *J. Clin. Endocrinol. Metab.,* **48**, 495–502.
Flier, J.S., Jarrett, D.B., Kahn, C.R. and Roth, J. (1976b), *Immunol. Commun.,* **5**, 361–373.
Flier, J.S., Kahn, C.R., Jarrett, D.B. and Roth, J. (1976a), *J. Clin. Invest.,* **58**, 1442–1449.
Flier, J.S., Kahn, C.R., Jarrett, D.B. and Roth, J. (1977), *J. Clin. Invest.,* **60**, 784–794.
Flier, J.S., Kahn, C.R., Roth, J. and Bar, R.S. (1975), *Science,* **190**, 63–65.
Harrison, L.C., Flier, J.S., Roth, J., Karlsson, F.A. and Kahn, C.R. (1979a), *J. Clin. Endocrinol. Metab.,* **48**, 59–65.
Harrison, L.C. and Itin, A. (1979), *Nature,* **279**, 334–336.
Harrison, L.C. and Kahn, C.R. (1980), in *Progress in Clinical Immunology,* Vol. 4 (Schwartz, R.S., Ed.), (in press).
Harrison, L.C., Martin, F.I.R. and Melick, R.A. (1976), *J. Clin. Invest.,* **58**, 1435–1441.
Harrison, L.C., Muggeo, M., Bar, R.S., Flier, J.S., Waldmann, T. and Roth, J. (1979b), *Clin. Res.,* **27**, 252A.
Harrison, L.C., Van Obberghen, E., Grunfeld, C., King, G.L. and Kahn, C.R. (1980), in *Membranes, Receptors and the Immune Response,* Alan R. Liss, Inc., New York, pp. 109–126.
Kahn, C.R., Baird, K.L., Flier, J.S. and Jarrett, D.B. (1977), *J. Clin. Invest.,* **60** 1094–1106.
Kahn, C.R., Flier, J.S., Bar, R.S., Archer, J.A., Gorden, P., Martin, M.M. and Roth, J. (1976), *N. Engl. J. Med.,* **294**, 739–745.
Kahn, C.R., Goldfine, I.D., Neville, D.M., Jr. and De Meyts, P. (1978), *Endocrinology,* **103**, 1054–1066.

Kahn, C.R. and Harrison, L.C. (1980), in *Carbohydrate Metabolism and its Disorders* (Randle, P., Steiner, D. and Whelan, W., eds), Academic Press, London (in press).

Karlsson, F.A., Van Obberghen, E., Grunfeld, C. and Kahn, C.R. (1979), *Proc. Natl. Acad. Sci. USA*, **76**, 809–813.

Kasuga, M., Akanuma, Y., Tsuchima, T., Suzuki, K., Kosaka, K. and Kibata, M. (1978), *J. Clin. Endocrinol. Metab.*, **47**, 66–77.

Kobayashi, M., Olefsky, J.M., Elders, J., Mako, M.E., Given, B.D., Schedwie, H.K., Fiser, R.H., Hintz, R.L., Horner, J.A. and Rubenstein, A.H. (1978), *Proc. Natl. Acad. Sci. USA*, **75**, 3469–3473.

Lawrence, J.C. Jr., Larner, J., Kahn, C.R. and Roth, J. (1978), *Mol. Cell. Biochem.*, **22**, 153–158.

LeMarchand-Brustel, Y., Gorden, P., Flier, J.S., Kahn, C.R. and Freychet, P. (1978), *Diabetologia*, **14**, 311–318.

Lippe, B.M., Golden, M.P., Kaplan, S.A. and Scott, M. (1979), *Pediatr. Res.*, **13**, 478.

Muggeo, M., Bar, R.S., Roth, J. and Kahn, C.R. (1979a), *J. Clin. Endocrinol. Metab.*, **48**, 17–25.

Muggeo, M., Kahn, C.R., Bar, R.S., Rechler, M.M., Flier, J.S. and Roth, J. (1979b), *J. Clin. Endocrinol. Metab.*, **49**, 110–119.

Muggeo, M., Flier, J.S., Abrams, R.A., Harrison, L.C., Deisserroth, A.B. and Kahn, C.R. (1979c), *N. Engl. J. Med.*, **300**, 477–480.

Olefsky, J.M. (1975), *J. Clin. Invest.*, **56**, 1499–1508.

Olefksy, J.M. (1976), *J. Clin. Invest.*, **57**, 1165–1172.

Olefsky, J.M. and Reaven, G.M. (1974), *J. Clin. Invest.*, **54**, 1323–1328.

Olefksy, J.M. and Reaven, G.M. (1977), *Diabetes*, **26**, 680–688.

Olefsky, J.M., Johnson, J., Liu, F., Jen, P. and Reaven, G.M. (1975), *Metabolism*, **24**, 517–527.

Roth, J. (1979), in *Endocrinology*, Vol. 3 (De Groot, L., Ed.), Grune and Stratton, New York, pp. 2037–2054.

Roth, J., Lesniak, M.A., Bar, R.S., Muggeo, M., Megyesi, K., Harrison, L.C., Flier, J.S., Wachslicht-Rodbard, H. and Gorden, P. (1979), *Proc. Soc. Exp. Biol, Med.*, **162**, 3–12.

Schilling, E.E., Rechler, M.M., Grunfeld, C. and Rosenberg, A.M. (1979), *Proc. Natl. Acad. Sci. USA*, **76**, 5877–5881.

Taylor, S.I., Podskalny, J.M., Samuels, B., Roth, J., Brasel, D.E., Pokora, T. and Engel, R.R. (1980), *Clin. Res.*, **28**, 408A.

Van Obberghen, E., Spooner, P.M., Kahn, C.R., Chernick, S.S., Garrison, M.M., Karlsson, F.A. and Grunfeld, C. (1979), *Nature*, **280**, 500–502.

Wachslicht-Rodbard, H., Muggeo, M., Saviolakis, G.A., Harrison, L.C., Flier, J.S. and Kahn, C.R. (1979a), *Clin. Res.*, **27**, 379A.

Wachslicht-Rodbard, H., Gross, H.A., Rodbard, D., Ebert, M.M. and Roth, J. (1979b), *N. Engl. J. Med.*, **300**, 882–887.

Yalow, R.S. and Berson, S.A. (1960), *J. Clin. Invest.*, **39**, 1157–1175.

8 Myasthenia Gravis and the Nicotinic Cholinergic Receptor

JON LINDSTROM and ANDREW ENGEL

Abbreviations

ACh	acetylcholine
AChE	acetylcholinesterase
AChR	acetylcholine receptor
MG	myasthenia gravis
EAMG	experimental autoimmune myasthenia gravis
αBGT	α-bungarotoxin
SDS	sodium dodecyl sulfate
epp	end-plate potential
mepp	miniature end-plate potential

Receptor Regulation

(*Receptors and Recognition*, Series B, Volume 13)

Edited by R. J. Lefkowitz

Published in 1981 by Chapman and Hall, 11 New Fetter Lane, London EC4P 4EE

Myasthenia gravis (MG) is a disease in which muscular weakness is caused by an autoimmune response to acetylcholine receptors (AChR's) in muscle. Neither the cause of this autoimmune response, nor a specific treatment for curing MG is known. However, a great deal has been learned about the pathological mechanisms by which the autoimmune response to AChR impairs neuromuscular transmission, thereby causing muscular weakness. Much of our understanding of these mechanisms has come from the study of animals with experimental autoimmune myasthenia gravis (EAMG) caused by immunization with AChR purified from fish electric organs. As a result of these studies we know that in MG and chronic EAMG the autoimmune response to AChR is mediated primarily by antibodies, rather than directly by lymphocytes. The primary effect of these antibodies is not to impair transmission by acting as simple competitive antagonists of acetylcholine binding, nor is the primary effect of antibodies to act indirectly as allosteric inhibitors of AChR function. Instead, the primary effect of anti-AChR antibodies is to reduce the amount of effective AChR. Antibodies cross-linking AChR in the post-synaptic membrane increase the rate of AChR destruction, causing a net decrease in AChR amount. Also, antibodies bound to AChR trigger the deposition of complement, which disrupts the postsynaptic membrane both by causing loss of AChR and by disrupting the ultrastructural localization of the AChR which remains.

Determining the structure of the AChR molecule and determining how this structure produces its function are of great interest for several reasons. As indicated above, the AChR is of interest as an antigen in an autoimmune disease. Of course, interest in AChR long predated interest in MG because it was the first neurotransmitter receptor recognized, and because the AChR now is far the best characterized receptor. This makes it something of an archetype for studies of other less accessible receptors. Pharmacological and electrophysiological studies of AChR have been both extensive, and in some cases quite elegant. But these approaches only reveal the functional properties of the molecule, while the structure responsible for producing these functions until recently has remained quite abstract. Now it is known that the AChR molecule is composed of four kinds of glycoprotein subunits. It is known that the α subunits participate in acetylcholine binding, and that the cation channel whose opening is regulated by acetylcholine binding is located within the AChR monomer. But we do not know the functions of the β, γ or δ subunits or the localization of these subunits within the molecule. There are, of course, many other things which we do not yet know about AChR structure, function and metabolism.

163

Several gifts of nature have contributed to rapid progress in studies of AChR structure, metabolism, and its role in MG and EAMG. Toxins from snake venoms which bind with high affinity and specificity to the acetylcholine binding site have proven ideal reagents for localizing, quantitating, and purifying AChR. Electric organs of fish have proven ideal sources of relatively large amounts of AChR for biochemical studies. Antibodies to AChR have proven excellent templates for comparing the structure of AChR from electric organs and muscle. Binding of these antibodies to AChR is so specific that antibodies to electric organ AChR cross react with muscle AChR by only a few per cent, but the basic structure of AChR from electric organ and muscle is so similar that antibodies to each of the four subunits of electric organ AChR can be used to recognize similar structures in muscle AChR. It is also the underlying similarity in structure between AChR of electric organ and muscle, of course, which is responsible for the development of EAMG in animals immunized with electric organ AChR.

In this review we will focus primarily on studies by our laboratories. In keeping with the theme of this volume on receptor regulation, we will start with a brief review of the structure of AChR molecule and the normal regulation of its metabolism, review the basic anatomy and physiology of the neuromuscular junction, and then proceed to consider how the immune response to AChR in EAMG and MG interferes with regulation of AChR amount, localization, and function at the neuromuscular junction.

8.2 AChR FUNCTION

The AChR is an integral protein of the postsynaptic membrane which transiently binds acetylcholine released by the nerve. Binding of acetylcholine triggers the transient opening of a rather non-specific (Huang *et al.*, 1978) channel (Katz and Miledi, 1972) through which cations, principally Na^+ and K^+, passively flow. In marine rays like *Torpedo californica*, the current flowing through AChR at thousands of synapses per electric organ cell (Heuser and Salpeter, 1979) sums over the millions of oriented cells to produce the electric discharge of the organ (Bennett, 1970). In mammalian muscle, the current flow through AChR at the single synapse per muscle fiber depolarizes the membrane, and if sufficient current flow occurs, voltage-sensitive channels outside of the end-plate trigger an action potential that propagates along the fiber and stimulates contraction of the fiber (Katz, 1966). In muscle from an animal with EAMG (Seybold *et al.*, 1976; Lambert *et al.*, 1976) or a person with MG (Desmedt and Borenstein, 1976; Elmquist *et al.*, 1964) the current flow through AChR at many end-plates is too small to trigger an action potential, so neuromuscular transmission is ineffective.

In rat muscle, binding of acetylcholine to a single AChR causes a square pulse of conductance 34 pS in amplitude and 1 ms in duration (Sakman, 1978). Two acetylcholine molecules appear much more effective than one in activating an AChR (Dionne *et al.*, 1978). At a human intercostal muscle end-plate, for example, the nerve impulse depolarizes the nerve ending, causing release by exocytosis (Heuser *et al.*, 1979) from the nerve ending of about 60 quanta of acetylcholine (Lambert and Elmquist, 1971), each of which contains about 10^4 acetylcholine molecules (Kuffler and Yoshikami, 1975). Much of this acetylcholine is lost to diffusion or acetylcholinesterase, but for every AChR activated about 5×10^4 ions flow across the postsynaptic membrane (Katz and Miledi, 1972). This provides substantial amplification of the current involved in depolarization of the small nerve ending, which is necessary in order to depolarize the much large muscle fiber. On continued exposure to agonists, AChR desensitizes. The ion channel of desensitized AChR is closed, and the binding affinity for agonists is greatly increased (Weber *et al.*, 1975; Sine and Taylor, 1979).

8.3 STRUCTURE OF *TORPEDO* ELECTRIC ORGAN AChR

AChR purified from the electric organs of the marine ray *T. californica* contains four subunits (Karlin *et al.*, 1976; Raftery *et al.*, 1976), as shown in Fig. 8.1. These subunits are present in the mole ratio $\alpha_2\beta\gamma\delta$ (Reynolds and Karlin, 1978; Damle and Karlin, 1978; Lindstrom *et al.*, 1979). All four are acidic glycoproteins (Lindstrom *et al.*, 1979a; Vandlen *et al.*, 1979) which are strongly non-covalently associated, but are dissociable by SDS. The apparent molecular weights of α, β, γ, and δ approximate 38, 50, 57 and 64×10^3 daltons respectively (Karlin *et al.*, 1976; Raftery *et al.*, 1976). There are two acetylcholine binding sites per AChR located on the α subunits (Reynolds and Karlin, 1978; Damle and Karlin, 1978). These sites are either functionally or structurally non-equivalent in that both can be labeled with $[^{125}I]$-α-bungarotoxin ($[^{125}I]$-αBGT), but only one can be labeled with agonist (Damle *et al.*, 1978) and antagonist (Damle and Karlin, 1978) affinity labeling reagents. *Torpedo* AChR normally exists as dimers formed by disulfide bonds between δ subunits (Chang and Bock, 1977; Hamilton *et al.*, 1977). Although the four AChR subunits have unique peptide maps (Lindstrom *et al.*, 1979a; Froehner and Rafto, 1979; Nathanson and Hall, 1979), and are basically immunochemically distinct (Claudio and Raftery, 1977; Lindstrom *et al.*, 1978; Lindstrom *et al.*, 1979b,c), there are some structural similarities. A monoclonal antibody to the δ subunit also reacts with the γ subunit, but with much lower affinity (Tzartos and Lindstrom, 1980). And a monoclonal antibody to the β subunit also reacts with the α subunit, but with much lower affinity (Tzartos and Lindstrom, 1980). The functions of the β, γ and δ subunits are unknown. It is known that the ion channel is an integral component

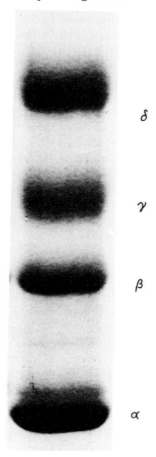

Fig. 8.1 Subunits of *Torpedo californica* AChR separated by electrophoresis in SDS on an acrylamide gel.

of the AChR monomer (Anholt *et al*., 1979), thus one or more subunit must contribute to its structure. When AChR is solubilized with the detergents Triton X-100 or sodium cholate, its ligand binding activity remains intact, but the ion channel is denatured (Lindstrom *et al*., 1980b). By solubilizing the AChR in mixtures of cholate and soybean lipid, the activity of the channel is preserved so that removal of the cholate by dialysis produces vesicles incorporating AChR in their membranes (Epstein and Racker, 1978). The AChR's in the reconstituted membrane respond to the binding of agonists by increasing the permeability of the vesicles to $^{22}Na^+$ (Epstein and Racker, 1978; Changeux *et al*., 1979; Wu and Raftery, 1979). It has recently become possible to purify AChR solubilized in cholate-lipid mixtures, and then re-incorporate the purified material into vesicles (Lindstrom *et al.*, 1980b; Huganir *et al.*, 1979).

By electron microscopy, fish electric organ AChR viewed from the end appear like lumpy doughnuts with negatively stained centers (Cartaud *et al.*, 1978; Allen and Potter, 1977; Rosss *et al.*, 1977). Pits are observed in the center of the surface of AChR molecules by freeze etching and in the center of AChR molecules split by freeze-fractures of AChR-rich membranes (Heuser and Salpeter, 1979). Tannic acid has also been reported to demon-strate a transmembrane channel through the center of the molecule (Potter and Smith, 1977). This suggests, but does not prove, that the ion channel runs through the center of the molecule. Electron microscopy (Heuser and Salpe-ter, 1979; Cartaud *et al.*, 1978; Allen and Potter, 1977; Ross *et al.*, 1977; Potter and Smith, 1977; Klymkowsky and Stroud, 1977; Karlin *et al.*, 1978; Stradler *et al.*, 1979), X-ray diffraction (Ross *et al.*, 1977), and neutron scat-tering (Wise *et al.*, 1979) provide data which are consistent with a model for AChR structure in which AChR viewed from the side appears somewhat mushroom-like, extending more on the extracellular than intracellular surface of the membrane. Fig. 8.2 shows a model of the AChR molecule which depicts some of its known and imagined structural features. *In vivo* cross-reaction of anti-subunit sera with AChR in muscle (Lindstrom *et al.*, 1978a) and binding of anti-subunit sera to vesicles of *Torpedo* electric organ mem-brane known to be oriented right side out shows that parts of all four subunits are located on the exterior surface of the membrane. Cytoplasmic filaments within the cell and basal laminae outside the cell may be associated with aggregations of AChR (Heuser and Salpeter, 1979; Jacob and Wentz, 1979).

8.4 SIMILARITIES BETWEEN THE STRUCTURE OF AChR FROM ELECTRIC ORGAN AND MUSCLE

AChR from the electric organs of the freshwater teleost *Electrophorus elec-tricus* have a subunit structure similar to that of AChR from *T. californica*, but this was recognized only recently (Lindstrom *et al.*, 1979c). Initially, two subunits were recognized in *Electrophorus* AChR (Lindstrom and Patrick, 1974), later three (Karlin *et al.*, 1976), and recently four (Lindstrom *et al.*, 1979c). The fourth subunit recognized, δ, is observed when the AChR is purified in the presence of iodoacetamide. Using both conventional antisera to *Torpedo* subunits and monoclonal antibodies, it can be shown that these correspond immunochemically to the α, β, γ and δ subunits of *Torpedo* AChR. The role of the iodoacetamide is unclear, but it may inhibit a protease which otherwise cleaves the δ subunit in several places so that when the AChR is dissociated in SDS and electrophoresed on polyacrylamide gels, the

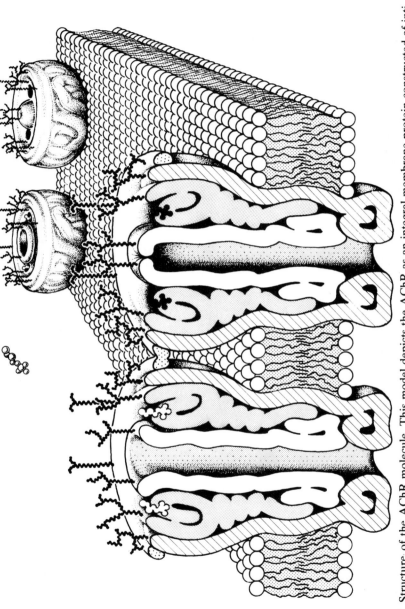

Fig. 8.2 Structure of the AChR molecule. This model depicts the AChR as an integral membrane protein constructed of intimately associated polypeptide subunits. Each AChR is thought to contain 2 α subunits and one each of β, γ, and δ. Most of the protein extends on the extracellular surface of the lipid bilayer. Carbohydrate-bearing portions of each subunit are exposed on the extracellular surface, as are the two acetylcholine binding sites located on α subunits. A transmembrane channel extends through the center of the molecule and its opening and closing is depicted as being produced by a small conformation change caused by binding of two acetylcholine molecules. The two AChR monomers sectioned in the foreground are depicted as joined into a dimer via a disulfide bond between their δ subunits. Reproduced by permission from Lindstrom (1979).

δ subunit is not apparent. Proteases can be a problem when studying AChR structure. For example, very mild trypsinization of *Electrophorus* AChR causes it to appear as a single band on polyacrylamide gels of slightly lower molecular weight than α (Lindstrom *et al.*, 1976a). Because α is much more protease resistant than β, γ or δ, very mild papain treatment of *Torpedo* AChR causes it to appear to be composed of only α. However, the proteolytically nicked β, γ and δ subunits remain associated with α and the AChR remains functional (Lindstrom *et al.*, 1980). These observations may explain the erroneous reports that AChR from *Torpedo* (Sobel *et al.*, 1977) or muscle (Shorr *et al.*, 1978) consists of only α subunits. In this context of considering artifacts, it is useful to remember the previously mentioned example of *Torpedo* AChR ion-channel function. The use of EDTA and iodoacetamide to inhibit Ca^{2+} and thiol-dependent proteases permits purification of AChR with all four subunits free of proteolytic degradation. However, if Triton X-100 or sodium cholate is used to solubilize the AChR, the ion channel will nonetheless be denatured (Lindstrom *et al.*, 1979d).

Although an elasmobranch like *Torpedo* and a teleost like *Electrophorus* are widely separated by evolution, the subunit structure of their AChR is similar in many respects. These similarities are: (1) close apparent molecular weight of four corresponding subunits [38, 50, 57 and 64 $\times$ 10^3 for *Torpedo* and 41, 50, 55 and 62 $\times$ 10^3 for *Electrophorus* α, β, γ and δ subunits (Karlin *et al.*, 1976; Raftery *et al.*, 1976; Lindstrom *et al.*, 1980)]; (2) presence of carbohydrate on all subunits (Vandlen *et al.*, 1979; Lindstrom *et al.*, 1979c); (3) specific affinity labeling of α subunits by a reagent that depends on a disulfide bond located at one end of the binding site (Karlin *et al.*, 1976); (4) immunochemical cross-reaction of α subunits (Lindstrom *et al.*, 1979b, 1980); (5) immunochemical cross-reaction of β subunits (Lindstrom *et al.*, 1979b, 1980); (6) spontaneous proteolytic degradation pattern of β subunits (Lindstrom *et al.*, 1979b, 1980); (7) immunochemical cross-reaction of γ subunits (Lindstrom *et al.*, 1979b, 1980); (8) immunochemical cross-reaction of δ subunits (Lindstrom *et al.*, 1979b, 1980); and (9) cross-reaction of an anti-δ monoclonal antibody with γ subunit (Lindstrom *et al.*, 1979b, 1980).

AChR purified from rat muscle has been reported to contain five subunits, termed $\alpha_1\alpha_2\beta\gamma\delta$ (Nathanson and Hall, 1979). α_1 and α_2 are both labeled with the affinity labeling reagent MBTA, and the amount of [^{3}H]MBTA bound is equal to the amount of [^{125}I]αBGT bound (Weinberg and Hall, 1979). This contrasts with the case of *Torpedo* and *Electrophorus* AChR where the two α subunits in the AChR monomer are indistinguishable by molecular weight, both bind [^{125}I]αBGT, and one binds [^{3}H]MBTA (Karlin *et al.*, 1976) with much greater affinity than the other (Wolosin *et al.*, 1980). Only a single size of [^{3}H]MBTA labeled α chain is observed in AChR from fetal calf muscle (Lindstrom *et al.*, 1979b). Despite differences in detail, these results clearly show that AChR from muscle has α-like subunits. In addition to α chains, AChR purified from rat muscle contains polypeptide chains similar in molecular weight to β, γ and δ subunits (Nathanson and Hall, 1979). Experiments using unpurified AChR from human (Lindstrom *et al.*, 1978d) and fetal calf muscle show that each contains four sets of antigenic determinants corresponding to α, β, γ and δ of *Torpedo* AChR, as shown in Fig. 8.3. It has not yet been shown that each set of antigenic determinants corresponds to a subunit, as in the case of *Electrophorus* AChR (Lindstrom *et al.*, 1980), but it

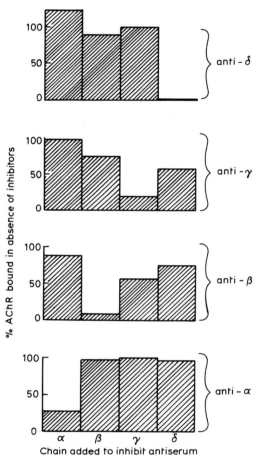

Fig. 8.3 Evidence for subunits in AChR from human muscle which are analogous to those in *Torpedo* electric organ AChR. Detergent-solubilized human AChR was labeled with [^{125}I]-α-bungarotoxin. Antisera to subunits of *Torpedo* AChR were shown to cross react detectably with the labeled human AChR. Binding of the antisera to the human AChR was shown to be specific by inhibition of antibody binding by preincubation of each anti-subunit antibody only by the appropriate *Torpedo* subunit. Data replotted from Lindstrom *et al.* (1978a).

seems likely that this will be the case. Thus it seems reasonable to expect that most of the basic structural features of the AChR molecule are similar in AChR from fish electric organs and mammalian muscle.

8.5 LOCALIZATION, SYNTHESIS AND DESTRUCTION OF AChR IN MUSCLE

AChR synthesis and destruction has been studied has been studied primarily using cell cultures. This very nice series of studies has recently been reviewed by Fambrough (1979) (Fambrough *et al.*, 1978). Muscle cells in culture synthesize AChRs, presumably on rough endoplasmic reticulum, and assemble them, to the point where they can bind [^{125}I]αBGT in less than 15 minutes. Completion of their assembly appears to require glycosylation. AChRs capable of binding αBGT are first observed in the Golgi apparatus oriented with their ligand binding sites toward the interior of membrane vesicles. Two to four hours are required before AChRs are finally incorporated into the muscle surface membrane, presumably by fusion of vesicles of newly synthesized AChRs with the surface so that the the AChRs are now oriented right side out. In cell culture, AChRs are localized in large and small patches all over the surface of the cell. Turnover of AChR in muscle cells in culture is very rapid. The half-time for degradation is around 20 hours. AChRs are selected randomly for degradation. The degradation process appears to involve endocytosis, which requires energy and involves cytoskeletal fibrils. The internalized AChRs are degraded in secondary lysosomes to their component amino acids which are then released from the cell. Degradation is usually measured by pre-labeling AChR on the cell surface with [^{125}I]-αBGT and then observing release of [^{125}I]tyrosine from the cells, which is thought to be produced by degradation of the [^{125}I]-αBGT simultaneously with the AChR to which it is attached. There is a lag of about 90 minutes between disappearance of AChR from the surface and the appearance of [^{125}I]tyrosine, which gives a measure of the time required for AChR degradation.

Before innervation of fetal muscle and after denervation of adult muscle, AChRs are localized diffusely over the surface of the muscle cell and turned over at the rapid rate characteristic of AChRs in muscle cells in culture ($t1/2 \simeq 17$ h) (Fambrough *et al.*, 1978; Fambrough, 1979). These are termed extrajunctional AChRs to distinguish them from the AChRs at neuromuscular junctions which are located at the tips of folds in the postsynaptic membrane (Fertuk and Salpeter, 1974) and turnover with the much slower half-time of about 150 hours (Heinemann *et al.*, 1978; Merlie *et al.*, 1979d). There is thought to be a small, but as yet undefined structural difference between junctional and extrajunctional AChRs (Nathanson and Hall, 1979). Antisera from patients with MG react better with extrajunctional AChR due to the presence of some determinants unique to extrajunctional AChR (Weinberg and Hall, 1979).

AChR metabolism is subject to regulation by a number of factors. Electrical stimulation of denervated muscles in organ cultures decreases the rate of AChR synthesis and destruction (Reiness and Hall, 1977). This is consistent with the idea that muscle activity is one of the trophic factors by which innervation influences the localization, amount and turnover of AChR. Hypophysectomy also decreases the rate of both AChR synthesis and destruction (Reiness *et al*., 1977), showing that AChR metabolism is sensitive to hormonal influences. As will be discussed later, binding of antibodies to junctional and extrajunctional AChRs increases their rate of destruction 2–3-fold (Heinemann *et al*., 1978; Merlie *et al*., 1979a; Heinemann *et al*., 1977; Merlie *et al*., 1979b; Lindstrom and Einarson, 1979; Kao and Drachman, 1977; Drachman *et al*., 1978a,b; Appel *et al*., 1977). This process is termed antigenic modulation and seems to use the same lysosomal mechanisms involved in normal AChR degradation.

Changes in AChR metabolism are not the only form of regulatory flexibility at the neuromuscular junction. In MG, for example, the end-plate regions appear unusually dispersed on the surface of the muscle fiber, which may be the result of sprouting of nerve endings and new synapse formation (Bickerstaff and Woolf, 1960), a process for which there is direct ultrastructural evidence (Engel *et al*., 1979). Sprouting of synapses may account for the report that MG muscle contains about twice the normal amount of acetylcholine, and that KCl-induced release of acetylcholine is also greater (Molenaar *et al*., 1979; Cull-Candy *et al*., 1980). Also, in rats with EAMG or in which AChR is blocked by αBGT, the evoked release of acetylcholine is reported to be increased nearly twofold (Molenaar *et al*., 1979). These results, if confirmed, might suggest that there is a presynaptic adaptive mechanism which attempts to compensate for decreases in postsynaptic sensitivity. It is important to remember all the adaptive regulatory mechanisms actually or potentially available for preserving the integrity of neuromuscular transmission when considering pathological mechanisms in EAMG and MG. The pathological mechanisms are themselves complex. Coupled with complex adaptive mechanisms which can also vary between individuals in an outbred population like man, it becomes clear that it is highly simplistic to expect, for example, that serum anti-AChR antibody concentration will exactly predict the severity of muscle weakness in MG. However, the complexity of the pathological versus adaptive mechanisms in EAMG or MG does not preclude understanding fundamental components of these processes.

8.6 ANATOMY AND PHYSIOLOGY OF THE NORMAL NEUROMUSCULAR JUNCTION

8.6.1 The fine structure of the motor end-plate

In order to understand the pathogenesis of myasthenia gravis, it is essential to understand the structure and function of the normal motor end-plate. For descriptive purposes, one can consider the motor end-plate as consisting of a number of regions, with each region made up of one terminal nerve arborization (or nerve terminal) and a postsynaptic region which underlies that arborization. At the end-plate region the Schwann cell covers only that part of the nerve terminal which faces away from the synaptic space.

Fig. 8.4 demonstrates the ultrastructure of a normal end-plate region. The nerve terminal contains synaptic vesicles, mitochondria and a varying

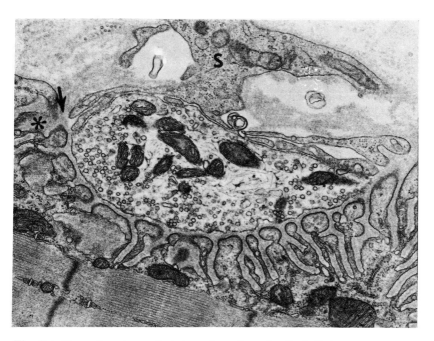

Fig. 8.4 Normal motor end-plate region of rat forelimb digit extensor muscle. Nerve terminal contains numerous synaptic vesicles, rare giant vesicles, sparse tubules and fine filaments, and several mitochondria. Junctional sarcoplasm (lower right) displays scattered ribosomes, rough endoplasmic reticulum, glycogen granules and mitochondria. Arrow and asterisk indicate primary and secondary synaptic cleft, respectively. S: Schwann cell. × 21 800. Reproduced by permission from Engel *et al.* (1976a).

complement of neurofilaments, microtubules, glycogen particles and lysosomal structures. On the average, mitochondria occupy 15 per cent of the nerve terminal volume and electron micrographs show 50 to 80 synaptic vesicles per μm^2 of the nerve terminal area. The synaptic vesicles have a mean diameter of 56 nm but a few vesicles are much larger (Engel and Santa, 1971; Engel *et al.*, 1975). Occasional vesicles contain dense cores and a few are coated with fuzzy material. At the frog motor end-plate those synaptic vesicles near the presynaptic membrane form clusters at regular intervals (Hubbard, 1973), but this is not seen at the human end-plate (Engel *et al.*, 1975). Freeze-fracture studies of the presynaptic membrane reveal parallel double rows of 10 nm integral membrane particles recurring at regular intervals (Heuser *et al.*, 1979). The particle rows are present in amphibians as well as mammals but in the latter their orientation is such that they overlie the shortest axis of the secondary clefts. Because at the amphibian end-plate the synaptic vesicles tend to cluster near the rows of particles, it has been inferred that the particles are related to the process which facilitates exocytosis of the synaptic vesicles and hence transmitter release. Recent freeze-fracture electron-microscopic studies of the ultra-fast frozen end-plate show this nicley (Heuser *et al.*, 1979).

In human intercostal muscle the junctional folds vary in height and complexity from end-plate to end-plate and even within the same end-plate (Engel and Santa, 1970). In the rat diaphragm, the junctional folds are more elongated at end-plates of 'white' than 'red' fibers (Padyulka and Gauthier, 1970) but in rat gastrocnemius and soleus muscles the differences are less marked (Santa and Engel, 1973). In human limb and intercostal muscles, the complexity of the postsynaptic regions varies somewhat from muscle to muscle (Engel *et al.*, 1975) but in a given muscle there are no conspicuous variations according to fiber types. In normal human external intercostal muscle, the postsynaptic membrane is approximately ten times longer than the presynaptic membrane (Engel *et al.*, 1975). The typical, normal junctional fold has a slender stalk and a terminal expansion or crest, which abuts on the primary synaptic cleft. Freeze-fracture electron microscopy of the junctional folds reveal rows of irregularly arrayed 8–12 nm (Heuser *et al.*, 1974) or 11–14 nm (Ellisman *et al.*, 1976) integral membrane particles at frog and rat end-plates. These are concentrated on the top 25 per cent of the folds, and are analogous to the AChR particles observed at the electroplaque (Heuser and Salpeter, 1979).

8.6.2 The relationship between acetylcholine and the synaptic vesicles

The synaptic vesicles contain quanta of acetylcholine. This is shown by the fact that stimulation of the nerve terminal in the presence of hemicholinium (Jones and Kwanbunbumpen, 1970) or depolarization of the nerve terminal with lanthanum (Heuser and Miledi, 1971) or the facilitation of the spontaneous release of transmitter quanta by black widow spider venom (Clark *et al.*, 1972) or prolonged stimulation at low frequency (Heuser and Reese, 1973) deplete the synaptic vesicles from the nerve terminals. On the other hand, the mechanism for packaging a relatively constant number of acetylcholine molecules into each vesicle is still not understood.

The average number of acetylcholine molecules in a single synaptic vesicle, or quantum, can be estimated by measuring the amount of transmitter released during repetitive stimulation, provided that acetylcholinesterase (AChE) is inhibited, and provided that the number of quanta released per nerve impulse and the number of end-plates in the preparation are known. The number of quanta released per impulse decreases as the stimulation frequency increases (Hubbard, 1973). For the rat diaphragm at 1 Hz stimulation there are about 8000 transmitter molecules per quantum (Kelly *et al.*, 1978). Another approach assesses the number of acetylcholine molecules needed to produce a quantal conductance change when delivered iontophoretically to an optimally exposed postsynaptic region (Kuffler and Yoshikami, 1975). For the garter snake end-plate this amounts to 6000 transmitter molecules. A vesicle with an internal diameter of 50 nm and iso-osmolar with its surroundings could accommodate about 6000 transmitter molecules.

8.6.3 AChR localization at the end-plate

More precise ultrastructural localization of AChR than is possible by radioautography or freeze-fracture electron microscopy can be achieved with

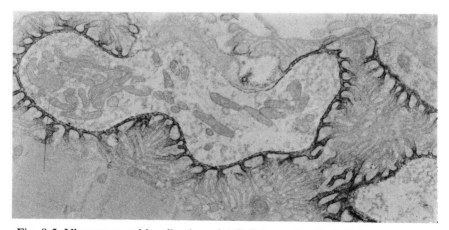

Fig. 8.5 Ultrastructural localization of AChR in rat forelimb digit extensor muscle with peroxidase-labelled α-bungarotoxin. AChR is associated with terminal expansions of postsynaptic folds. Presynaptic membrane is also stained. × 13 100. Reproduced by permission from Engel *et al.* (1977a).

peroxidase-labeled α-bungarotoxin (Engel *et al.*, 1977). This clearly shows that the AChR is concentrated on the terminal expansions of the junctional folds (Fig. 8.5). Morphometric analysis of electron micrographs indicates that at the normal human external intercostal end-plate 30 per cent of the post-synaptic membrane reacts for AChR (Engel *et al.*, 1977).

The presynaptic membrane also reacts for AChR, though less intensely than the postsynaptic membrane. It is not yet certain that the presynaptic localization is genuine, and it is likely that the labeled toxin here binds to a molecule which has the same structure, physiological properties and antigenic determinants as the postsynaptic AChR. Regardless of their exact significance, presynaptic toxin binding sites cannot account for more than 10 per cent of the total toxin binding sites at the end-plate (Porter and Barnard, 1975).

8.6.4 The end-plate acetylcholinesterase (AChE)

Unlike AChR, AChE is associated with the entire postsynaptic membrane and is not an integral membrane protein (Rosenberry, 1975). AChE is associated with the basement membrane which surrounds the junctional folds and can be dissociated from the end-plate by proteolytic digestion (McMahan *et al.*, 1978). In normal muscle, 40 per cent of the total AChE is associated with the end-plate and the remainder is distributed diffusely along the rest of the muscle fiber (Hall, 1973). Different molecular forms of AChE, composed of varying numbers of subunits and having different sedimentation properties, exist.

8.6.5 Postsynaptic events after the release of a single quantum from the nerve terminal

In the resting state there is a steady, but random, release of transmitter quanta from the nerve terminal. The frequency of release increases with depolarization of the nerve terminal, temperature, the ambient calcium concentration and with stretching of the nerve terminal (Hubbard, 1973; Martin, 1977).

The release of a single quantum leads to the following events: (1) diffusion of ACh molecules from the site of release into the synaptic space; (2) collisions between ACh molecules and AChRs; (3) binding of ACh molecules to AChRs; (4) ionic conductance changes induced by ACh–AChR interactions; (5) dissociation of ACh molecules from AChRs; (6) hydrolysis of ACh to choline and acetate by AChE and (7) retrieval of the products of hydrolysis by the nerve terminal. Collisions between ACh and AChR are stochastic events. Some ACh molecules are hydrolysed by AChE or escape from the synaptic space before they can collide with AChR, and only some of the collisions result in attachment of ACh to a specific binding site on AChR. At the normal end-plate, the strategic deployment of the AChR on the crests of the junctional folds, 70–100 nm from the release sites, assures exposure of AChR to a relatively high concentration of ACh. In addition, the high initial concentration of ACh hinders the action of AChE by substrate inhibition (Fertuk and Salpeter, 1976). These factors favor the occurrence of most of the collisions between ACh and AChR immediately after release and are responsible for the short rise time of the quantal conductance change. Subsequently, as unbound ACh diffuses laterally into the primary and radially into the secondary clefts, its concentration diminishes and its hydrolysis by AChE increases. The depths of the secondary clefts, with sparse AChR and abundant AChE, serve as cul de sacs in which unbound ACh is trapped and hydrolysed and from which choline can diffuse back into the nerve terminal.

Under normal conditions and with AChE fully active, ACh molecules in a single quantum exert their effect within a radius of 0.8 μm from the site of release (Hartzell *et al.*, 1975). Spread of the transmitter molecules over that distance in the primary cleft would expose them to 1.5–2 μm^2 of postsynaptic membrane over the crests of the folds and hence to about 30 000–40 000 ACh binding sites. From this it is clear that a single quantum containing 6000–8000 molecules of ACh does not saturate all postsynaptic receptors within the area of its spread. AChE limits the number of collisions of ACh and AChR, the radius of spread of ACh and the duration of the quantal conductance change. Anticholinesterase drugs increase the lateral spread of the transmitter, the number of collisions with AChR and the duration of the quantal conductance change (Hartzell *et al.*, 1975; Katz and Miledi, 1973).

Interaction of ACh with AChR increases the permeability of the postsynaptic membrane to cations but not anions. Sodium, potassium, calcium and magnesium conductances all increase, and movement of these ions occurs along their electrochemical gradients. Since the gradient for sodium is the steepest and its conductance exceeds that of potassium, the bulk of the quantal·current is due to a net influx of sodium ions (Takeuchi and Takeuchi, 1960).

The quantal conductance change is the sum of elementary conductance changes occurring at individual ion gates. The opening of a gate depends on the binding of one, or more probably two, ACh and AChR. Parameters of the elementary event can be studied by noise analysis of the current which flows across the end-plate when ACh is applied to it electrophoretically (Katz and Miledi, 1972; Anderson and Stevens, 1973). For the frog end-plate at 20° C, Katz and Miledi (1972) found that the elementary event had an amplitude of 0.3 μV, a duration of 1 ms and that the conductance change was 100 pmhos. Anderson and Stevens (1973) obtained 32 pmhos for the conductance change and 2–3 ms for the duration of the event. If the elementary event had an amplitude of 0.3 μV and the mepp amplitude was 0.6 mV, one quantum would open 2000 ion gates. Assuming that one quantum contains 6000 ACh molecules, that two ACh molecules open a gate and that each ACh molecule attaches to the receptor but once before hydrolysis, two-thirds of the ACh molecules in the quantum would have to contribute to the mepp response. If these assumptions are correct, relatively small decreases in the number of transmitter molecules in the quantum will decrease the mepp amplitude even when postsynaptic mechanisms of transmission are normal.

There is evidence that the duration of the elementary event corresponds to the occupation time of the receptor by ACh (Katz and Miledi, 1973). Prostigmine has no effect on the duration of the elementary event but prolongs the duration of the quantal conductance change. This can be explained by assuming that when AChE is inhibited, single molecules of ACh pass from receptor to receptor, becoming attached to each for about one ms, before escaping from the synaptic space. D-Tubocurarine has no effect on the duration or amplitude of the elementary event but decreases the amplitude of the mepp (Katz and Miledi, 1972). This is consistent with D-tubocurarine competing with ACh for binding to AChR.

The amplitude of the mepp depends not only on the number of elementary events but also on the input resistance and the resting membrane potential of the muscle fiber (Katz and Thesleff, 1957). For a resting membrane potential of 80 mV, the mean mepp amplitude at the normal, adult, human external intercostal end-plate is approximately 0.9 mV (Lambert and Elmquist, 1971).

8.6.6 Postsynaptic events after depolarization of the nerve terminal by the nerve impulse

Depolarization of the nerve terminal by the nerve impulse results in a nearly synchronous release of a relatively large number of quanta from the nerve terminal. The number of quanta released depends, among other factors, on the frequency of stimulation (Elmquist and Quastel, 1965). One per second stimulation releases approximately 60 quanta per impulse at the human external intercostal muscle end-plate (Lambert and Elmquist, 1971) and about 180 quanta per impulse at the rat diaphragm end-plate (Lambert *et al.*, 1976). Under normal circumstances, when AChE limits lateral diffusion of ACh in the synaptic cleft, individual quanta depolarize non-overlapping regions of the postsynaptic membrane. The depolarization produced by the individual quanta summate, but not linearly (Martin, 1977) producing a larger depolarization, the end-plate potential (epp). When the epp exceeds a certain threshold, it triggers a self-regenerating muscle action potential.

The amplitude of the epp is affected by the same factors that affect the amplitude of the mepp and, in addition, by the number of quanta released by nerve impulse. The difference between the number of quanta released by nerve impulse and the number of quanta required to trigger the action potential represents the safety margin of neuromuscular transmission.

8.6.7 Presynaptic events after depolarization of the nerve terminal by the nerve impulse

Depolarization of the nerve terminal by the nerve impulse is followed by an influx of calcium into the nerve terminal and it is this influx that mediates transmitter release (Katz and Miledi, 1976a,b). The entry of calcium also results in presynaptic facilitation, so that the probability of release by a subsequent impulse is increased. The effects of calcium are antagonized by magnesium (Martin, 1977).

The number of quanta released by a nerve impulse (m), or the quantum content of the epp, depends on the probability of release (p) and on at least one additional factor designated as n. The basic law of the quantum theory of synaptic transmission states that $m = np$ (Del Castillo and Katz, 1954). The factor n was originally defined as the number of quantal units capable of responding to the nerve impulse (Del Castillo and Katz, 1954). Later it was assumed that n represented the number of quanta readily available for release (Elmquist and Quastel, 1965; Bennett and Florin, 1974) or, alternatively, the number of active release sites in the nerve terminal (Zucker, 1973; Wernig, 1975).

Repetitive stimulation of the end-plate under conditions of high initial quantum release (physiological calcium and magnesium levels) is associated with a frequency-dependent decrease of the epp amplitude to a certain plateau. The decline has been attributed to a decrease of m due to a decrease in the number of synaptic vesicles readily available for release (or n) while p remained constant; and the plateau to equilibrium between mobilization and release of quanta (Elmquist and Quastel, 1965). This might be an oversimplification, for a decrease in p could also contribute to the depression of the epp amplitude (Martin, 1977). Regardless of the exact mechanism of the decrease of the epp amplitude during repetitive stimulation, it is associated with a reduction of the safety margin of neuromuscular transmission. In myasthenia gravis, where the initial amplitude of the epp, and hence the safety margin of transmission, is reduced, the decline during repetitive stimulation results in an epp which is sub-threshold for triggering the muscle action potential. The stimulation-dependent failure of the epp to trigger the muscle action potential explains the decremental electromyogram and the abnormal fatigability during exertion which characterize the disease. These phenomena will be further dealt with in subsequent sections.

Repetitive stimulation under conditions of low quantum release (high magnesium or low calcium concentration) is associated with increased transmitter release (Hubbard, 1973; Martin, 1977). According to current knowledge, at least four physiologically distinct stimulus-dependent processes can increase transmitter release (Magleby, 1979; Glavinovic, 1979). (1) Very short-term facilitation appears immediately after a single stimulus and decays with a half-time of about 35 ms. (2) Short-term facilitation appears about

80 ms after a single stimulus, peaks within 40 ms and decays with a half-time of about 250 ms. (3) Augmentation becomes significant after repeated stimuli and decays with a half-time of about 7 seconds. (4) Post-tetanic facilitation increases in amplitude and duration with the number of stimuli and decays with a half-time of 20–60 seconds. The two processes of facilitation are major factors for increasing m during the first few impulses while augmentation and post-tetanic potentiation develop later. Although the different processes which increase transmitter release are best studied under conditions of low quantum release, they also occur under physiological conditions and tend to antagonize the process of depression. The temporal profiles of the opposing processes explain why the transmission defect in myasthenia gravis is most readily observed at low (2–3 Hz) frequencies of stimulation, the transient improvement of the defect a few seconds after tetanic stimulation, and the worsening of the defect one or two minutes after tetanic stimulation.

8.7 IMMUNIZATION WITH AChR

EAMG can be induced in all species tested by immunization with AChR purified from fish electric organs. The species tested have included rabbits (Patrick and Lindstrom, 1973; Aharonov *et al.*, 1975; Sugiyama *et al.*, 1973; Heilbronn *et al.*, 1975; Penn *et al.*, 1976; Berti *et al.*, 1976; Green *et al.*, 1975), rats (Lindstrom *et al.*, 1976; Lennon *et al.*, 1975), mice (Fuchs *et al.*, 1976; Fulpius *et al.*, 1976; Granato *et al.*, 1976; Christadoss *et al.*, 1979a; Berman and Patrick, 1980), guinea pigs (Lennon *et al.*, 1975; Fuchs *et al.*, 1976; Fulpius *et al.*, 1976; Granato *et al.*, 1976; Christadoss *et al.*, 1979a; Berman and Patrick, 1980; Tarrab-Hazdai *et al.*, 1975a), goats (Lindstrom, 1976; Lindstrom *et al.*, 1978b), monkeys (Tarrab-Hazdai *et al.*, 1975b), and frogs (Nastuk *et al.*, 1979). Detailed features of the muscle weakness characteristic of EAMG vary depending on the species immunized. For example, monkeys with EAMG exhibit the dropping eyelids characteristic of human MG (Tarrab-Hazdai *et al.*, 1975b), whereas rodents do not (Patrick and Lindstrom, 1973; Aharonov *et al.*, 1975; Sugiyama *et al.*, 1973; Heilbronn *et al.*, 1975; Penn *et al.*, 1976; Berti *et al.*, 1976; Green *et al.*, 1975; Lennon *et al.*, 1975; Fuchs *et al.*, 1976; Christadoss *et al.*, 1979a; Fulpius *et al.*, 1976; Granato *et al.*, 1976; Berman and Patrick, 1980; Tarrab-Hazdai *et al.*, 1975a). The sensitivity to EAMG and course of response varies, for example, between rabbits, rats and mice. Most rabbits rapidly become nearly moribund 25–30 days after immunization (Patrick and Lindstrom, 1973; Sugiyama *et al.*, 1973; Aharonov *et al.*, 1975; Heilbronn *et al.*, 1975; Green *et al.*, 1975; Penn *et al.*, 1976; Berti *et al.*, 1976), while strains of mice differ in their response and may require multiple immunizations over much longer periods to become weak (Fuchs *et al.*, 1976; Christados *et al.*, 1979a,b; Berman and

Patrick, 1980). Presumably these differences reflect differences in genetically determined immune responsiveness, in degree of cross-reaction of AChR, relative effectiveness of various components of the immune response, relative effectiveness of adaptive responses at the neuromuscular synapse, physiological differences in the safety margin of neuromuscular transmission, and anatomical differences. The best characterized animal for studying EAMG is the Lewis strain of rat.

EAMG has been induced in Lewis rats by immunization with AChR purified from syngenic animals (Lindstrom *et al.*, 1976b), fetal calf muscle (unpublished), electric eels (Lindstrom *et al.*, 1976a,b; Lennon *et al.*, 1975). *Torpedos* (Lindstrom and Einarson, 1979), and the purified subunits of *Torpedo* AChR (Lindstrom *et al.*, 1978a, 1979b). Native *Torpedo* AChR is a very potent immunogen and immunization with a single microgram in complete Freund's adjuvant produces measurable antibody and loss of muscle AChR (Lindstrom and Einarson, 1979). However, because of the low degree of cross-reaction with muscle AChR, single or multiple doses of 15–30 μg are usually used. Antisera to native AChR cross-react detectably, but not well, with SDS-denatured AChR subunits (Lindstrom *et al.*, 1978a, 1979b). Interspecies cross-reaction of antisera to native AChR is greatest with denatured α subunits (Lindstrom *et al.*, 1979b). *Torpedo* AChR subunits dissociated and purified in SDS are less immunogenic than the native molecule by several hundredfold (Lindstrom *et al.*, 1978a, 1979b). Antibodies to the denatured subunits react well with native AChR, but recognize antigenic determinants different from those which predominant on the native molecule. Thus, the most important antigenic determinants on AChR depend on the native quaternary configuration of the molecule. About fifty per cent of the antibodies directed at the native molecule bind at or near a determinant on the α subunit outside the acetylcholine binding site. This is shown by the observation that monoclonal antibodies to this site can inhibit the binding of about fifty per cent of the antibodies in antiserum to native AChR (Tzartos and Lindstrom, 1980). Monoclonal antibodies derived from hybridomas of immunized rat spleen cells with mouse myeloma cell lines should reflect statistically the composition of antibodies in antisera to AChR. And, in fact, most monoclonal antibodies derived from cloned hybridomas are directed at the α subunit (Tzartos and Lindstrom, 1980). Nonetheless, it should be remembered that muscle AChR also contains antigenic determinants corresponding to β, γ and δ (Nathanson and Hall, 1979; Lindstrom *et al.*, 1978a, 1979b), that immunization with α, β, γ or δ can produce EAMG (Lindstrom *et al.*, 1978a), and that MG patients produce antibodies directed at mammalian β and γ as well as α (Lindstrom, unpublished observation).

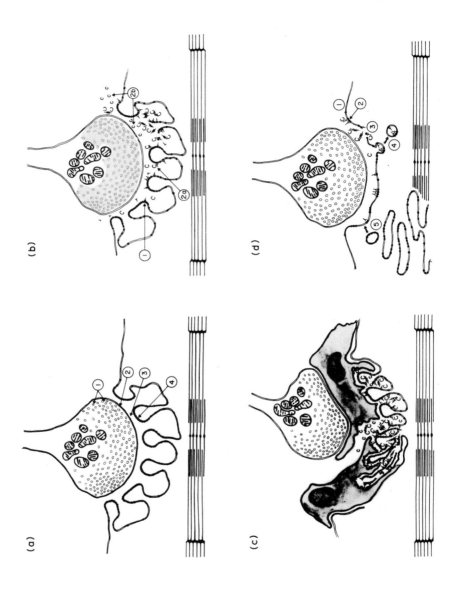

(a)

(b)

(c)

(d)

8.8 ACUTE, PASSIVE AND CHRONIC EAMG IN RATS

After a single immunization with AChR in complete Freund's adjuvant, plus injection of pertussis vaccine as additional adjuvant at other sites, Lewis rats undergo two successive phases of muscular weakness, an early acute phase and a later chronic phase (Lennon *et al*., 1975). Fig. 8.6 summarizes the features of these two phases. The acute phase of muscular weakness occurs 8–11 days after immunization (Lennon *et al*., 1975; Lindstrom *et al*., 1976b). At low doses of AChR this is mild, and if pertussis is not used as additional adjuvant, acute EAMG is not observed. Acute weakness lasts for two or three days, after which the rats appear normal until day 28–30 when chronic muscular weakness begins (Lennon *et al*., 1975; Lindstrom *et al*., 1976b). This weakness may be progressive until death (Lindstrom *et al*., 1976a), or diminish as the response to the immunogen diminishes (Lindstrom, 1977). Antisera from

---

Fig. 8.6 (opposite) The course of EAMG in rats (a) Depicts a normal neuromuscular junction showing (1) acetylcholine vesicles in the nerve ending, (2) the presynaptic membrane of the nerve ending, (3) the synaptic cleft between nerve and muscle, and (4) the postsynaptic membrane densely packed with AChR at the tips of the folds adjacent to the nerve ending. (b) Depicts a junction 6 or 7 days after immunizing a rat with electric organ AChR. (1) Antibodies are bound to a small fraction of the AChR, resulting in the fixation of complement (2a). Fixing of complement triggers a cascade of reactions that results in focal lysis of the postsynaptic membrane at the tips of the folds where AChR are concentrated and in the proteolytic release of fragments of complement which promote the migration of phagocytes (2b). (c) Depicts the phagocytic invasion of the end-plate observed in acute and passive EAMG. The phagocytes interact with bound antibody and bound C3 components of complement and phagocytize large areas of postsynaptic membrane containing more AChR than bound antibody. At many junctions the phagocytic invasion completely disrupts neuromuscular transmission resulting in transient functional denervation. (d) Depicts the simplified postsynaptic membrane characteristic of chronic EAMG and MG. No phagocytes are seen. The postsynaptic membrane (1) lacks its former complex folded structure. Most of the AChR have been lost and most of those which remain have antibodies bound (2). Complement fixed by the bound antibody causes focal lysis (3) which is probably responsible for destruction of the complex folded structure of the postsynaptic membrane. Antibodies crosslink AChR into aggregates which are endocytosed (4) into lysosomes and destroyed in the course of antigenic modulation. This increase in the rate of AChR destruction is partly responsible for the decrease in AChR content. It is unknown whether *in vivo* the rate of AChR synthesis (5) increases from the low rate characteristic of mature junctions to the high rate characteristic of denervated or fetal muscle in order to partially compensate for the increased rate of AChR destruction. Reproduced from Lindstrom (1979) by permission.

a rat with chronic EAMG can very efficiently passively transfer EAMG to a normal rat (Lindstrom *et al.*, 1976c). The recipient rat begins to exhibit weakness within 12–24 hours, but, if not fatal, this passes within a few days. EAMG can also be passively transferred with lymphocytes, but not very efficiently (Lennon *et al.*, 1976), and then the weakness is delayed by many days, presumably due to the time required for these lymphocytes to synthesize sufficient antibody to produce an effect.

Both acute and passive EAMG are characterized by a massive phagocytic invasion of end-plates (Engel *et al.*, 1976a,b, 1977a,b, 1979; Lindstrom *et al* 1976c; Sahashi *et al.*, 1978). These phagocytes are presumably attracted by chemotactic fragments released by activated complement and bind to antibody and complement deposited on the postsynaptic membrane. This is shown by the observations that the phagocytic invasion occurs after the injection of anti-AChR into a normal rat (Lindstrom *et al.*, 1979c; Engel *et al.*, 1979), but is inhibited if the normal rat is first depleted of the C3 component of complement (Lennon *et al.*, 1977). The phagocytes destroy the postsynaptic membrane producing functional denervation in many fibers (Lambert *et al.*, 1976). Only a small fraction of the AChR are labeled with antibody during acute or passive EAMG (Lindstrom *et al.*, 1976b,c), but this permits fixation of sufficient complement to target the postsynaptic membrane for destruction. Large loss of muscle AChR is observed during the phagocytic invasion (Lindstrom *et al.*, 1976b,c). After two or three days, when the phagocytes have left the muscle AChR content transiently increases to more than normal (Lindstrom *et al.*, 1976b,c), presumably due to formation of extrajunctional AChR in response to the transient denervation occurring in many fibers. The phagocytic response may be triggered by the very rapid deposition of complement after immunization using multiple adjuvants or passive transfer of antibody. Dynamics of complement deposition and inactivation may also be responsible for the rapid termination of this response. Chronic EAMG is observed even if pertussis is not used as additional adjuvant and no acute phase is seen. There is no equivalent of acute EAMG in human MG.

Chronic EAMG is characterized morphologically by the presence of a simplified postsynaptic membrane containing diminished amounts of AChR to which antibody and complement are bound (Engel *et al.*, 1976a,b, 1977a,b, 1979; Sahashi *et al.*, 1978). Phagocytes are not observed. But there is evidence of complement-mediated focal lysis of the postsynaptic membrane (Sahashi *et al.*, 1978). A decrementing electromyogram typical of MG is observed (Seybold *et al.*, 1976). Micro-electrophysiological studies reveal normal numbers of acetylcholine quanta released (Lambert *et al.*, 1976), but reduced acetylcholine sensitivity of the postsynaptic membrane (Bevan *et al.*, 1976), which accounts for a reduction in the size of miniature end-plate potentials to about one third of normal (Lambert *et al.*, 1976). The content of AChR in muscle is reduced to about one third of normal, and most of the

AChRs which remain have antibodies bound (Lindstrom and Einarson, 1979; Lindstrom *et al.*, 1976b).

AChR is a very potent immunogen, and immunization with small amounts causes substantial loss of AChR from muscle, as shown in Fig. 8.7. The

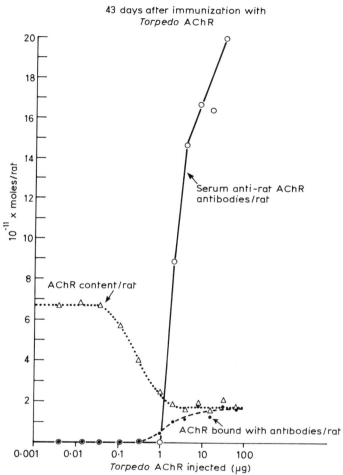

Fig. 8.7 Efficiency of immunization of rats with *Torpedo* AChR at inducing EAMG. At day 0, groups of four rats were injected intradermally with the indicated amounts of AChR emulsified in complete Freund's ajuvant. Then 43 days later all the rats were sacrificed and the amount of muscle AChR, serum antibodies capable of cross-reacting with rat muscle AChR, and antibody-bound muscle AChR were measured. Reproduced from Lindstrom and Einarson (1979) by permission.

muscular weakness characteristic of chronic EAMG can be readily explained by this decrease in AChR. Simple competitive inhibition of AChR with toxin produces very similar clinical and electrophysiological signs (Satyamurti *et al.*, 1975). However, several lines of evidence indicate that this is not the way antibody to AChR acts. Few or none of the antibodies are directed at the acetylcholine binding site (Patrick *et al.*, 1973; Lindstrom, 1976). Antisera can directly inhibit AChR function (Heinemann *et al.*, 1977; Bevan *et al.*, 1978), but in general the effects of antibody on AChR function are small. Rat muscle cells in culture exposed to anti-AChR and then subjected to acetylcholine noise analysis showed approximately a 23 per cent decrease in the open time and a 15 per cent decrease in the conductance of the ACh induced ion channel (Heinemann *et al.*, 1977) and cultured human muscle cells gave similar results (Bevan *et al.*, 1978). Because the safety factor for neuromuscular transmission is large, these effects would not in themselves inhibit transmission even if all AChR of the junctional were antibody bound. This is consistent with the observation that in normal rats depleted of the C3 component of complement by treatment with cobra venom factor, injected anti-AChR antibodies can bind at least 67 per cent of muscle AChR without impairing transmission (Lennon *et al.*, 1977).

Complemented-mediated destruction of the postsynaptic membrane subsequent to anti-AChR antibody binding appears to impair transmission in two ways. Focal lysis of the postsynaptic membrane releases membrane fragments containing AChR, antibody and complement (Sahashi *et al.*, 1978). Thus, lysis contributes directly to loss of AChR. The postsynaptic membrane must reseal very effectively after such lytic attacks, because the resting membrane potential in chronic EAMG is not significantly reduced (Lambert *et al.*, 1976). Complement-mediated lysis of the postsynaptic membrane is also responsible for the destruction of the terminal portions of the junctional folds and simplification of the postsynaptic region (Engel *et al.*, 1976a,b, 1977b, 1979). AChR is normally concentrated at the terminal expansions of the folds in a strategic location to interact with acetylcholine released from the nearby nerve terminal. Disruption of this spatial relationship may further impair transmission, but it is difficult to estimate the effects of altered end-plate geometry on neuromuscular transmission.

Antigenic modulation of AChR produces AChR loss by increasing the rate of AChR destruction, as shown in Fig. 8.8. Crosslinking of AChR by antibody (Drachman *et al.*, 1978b; Lindstrom and Einarson, 1979) causes AChR aggregation and internalization (Prives *et al.*, 1979), where they are degraded in lysosomes (Merlie *et al.*, 1979a,b). In cell culture the internalization process is energy, temperature and cytoplasmic-filament dependent, and requires about an hour and a half before amino acid residues are released from the cell (Heinemann *et al.*, 1977, 1978; Reiness and Hall, 1977; Reiness *et al.*, 1977; Appel *et al.*, 1977; Kao and Drachman, 1977; Drachman *et al.*, 1978a,b;

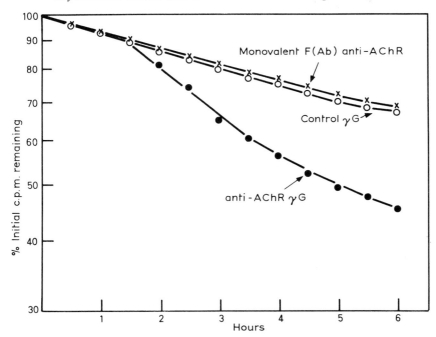

Fig. 8.8 Antigenic modulation of AChR crosslinked by antibodies. The AChR of muscle cells in culture were labeled with [^{125}I]-α-bungarotoxin, then the amount of [^{125}I]tyrosine released into the medium was determined. Degraded labeled AChR was measured at intervals and the amount of labeled AChR remaining on the cells calculated. Cells exposed to IgG from normal rat serum or monovalent anti-AChR F(Ab) destroyed their AChR at the same rate. But cells exposed to bivalent anti-AChR IgG destroyed their AChR at a faster rate. The rate of [^{125}I]tyrosine release did not increase until an hour and a half after the addition of antibody, due to the time required to crosslink AChR, endo-cytose, proteolyse, and release the amino acid residue. Reproduced from Lindstrom and Einarson (1979) by permission.

Merlie *et al.*, 1979a,b; Lindstrom and Einarson, 1979). Lysosomal protease inhibitors can prevent destruction of the internalized AChR (Merlie *et al.*, 1979a,b). In all these respects, except for the crosslinking and aggregation, the mechanism of antigenic modulation resembled the normal mechanism of AChR destruction. Antigenic modulation accelerates the rate of AChR de-struction by 2 to 3-fold in both cell culture (Heinemann *et al.*, 1977; Kao and Drachman, 1977; Drachman *et al.*, 1978a,b; Appel *et al.*, 1977; Lindstrom and Einarson, 1979) and organ culture (Heinemann *et al.*, 1978; Merlie *et al.*, 1979a). And muscle from rats with EAMG does, in fact, destroy its AChR at the accelerated rate expected of antigenic modulation (Merlie *et al.*, 1979b).

Endocytosis may be the rate-limiting step of antigenic modulation (Lindstrom and Einarson, 1979). A threefold increase in the rate of AChR destruction would be sufficient to account for the observation that AChR content in the muscles of rats with EAMG decreases to a minimum of about one third of normal (Lindstrom and Einarson, 1979). However, it is unknown whether *in vivo* there is an adaptive increase in the rate of AChR synthesis from the very low rate characteristic of normal junctional AChR toward the much higher rate characteristic of extrajunctional AChR. Thus, although it is clear that antigenic modulation is a very important mechanism for causing AChR loss and thereby impairing transmission, it is not yet possible to quantitate its effects with respect to those of antibody dependent, complement-mediated focal lysis.

No special anti-muscle AChR antibody specificity appears to be required to cause EAMG. As previously discussed, antibodies against the acetylcholine binding site are not required. Crosslinking of AChR by antibody to trigger antigenic modulation or simple binding of antibody to trigger fixation of complement should not depend on to which part of the AChR molecule the antibody binds. In fact, immunization with any of the four denatured AChR subunits causes EAMG (Lindstrom *et al.*, 1978a). However, differences in anti-AChR subclass could effect the ability to bind complement. And, in theory, some effects of antibody specificity should exist. For example, antibodies to determinants on the interior of the cell membrane would not be able to bind *in vivo*, and antibodies to determinants in the center of the molecule or represented twice on the molecule might not effectively crosslink AChR (because both antibody binding sites could bind within the monomer). On the other hand, antibodies to the acetylcholine binding sites or ion channel opening might be very effective at inhibiting AChR function. Such effects have yet to be demonstrated with monoclonal populations of anti-AChR antibodies, but if they are, then that should tell us as much about the AChR molecule as about the pathological mechanisms of EAMG.

8.9 MORPHOLOGICAL FEATURES AND IMMUNOPATHOLOGY OF THE NEUROMUSCULAR JUNCTION IN EAMG

8.9.1 Acute EAMG

The ultrastructure of the motor end-plate in acute and chronic EAMG of the rat was investigated in detail (Engel *et al.*, 1976a,b). During acute EAMG, between 7 and 11 days after immunization with AChR, a striking and rapidly evolving sequence of events takes place. The initial change is disintegration of terminal expansions of the junctional folds into globular residues (Fig. 8.9). Shortly after this, many postsynaptic regions split away from the underlying

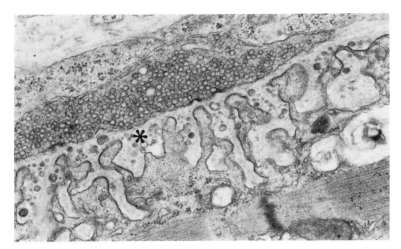

Fig. 8.9 Acute EAMG. Tips of numerous junctional folds are degenerating and globular fragments of folds are accumulating in widened synaptic space (asterisk. A small autophagic vacuole is present in junctional sarcoplasm on the right. × 27 000. Reproduced by permission from Engel *et al*. (1976a).

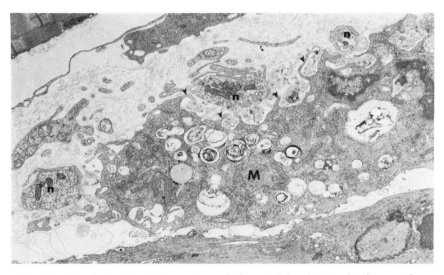

Fig. 8.10 Acute EAMG. Macrophage (M) containing heterophagic vacuoles separates nerve terminals (n) from underlying muscle fiber. Region of muscle fiber under macrophage shows degenerative changes. Space between nerve terminals and macrophage contains degraded residues of synaptic folds. Macrophage tentacles partially surround the degraded material (arrowheads). × 9500. Reproduced by permission from Engel *et al*. (1976b).

muscle fibers and the degenerating and separated junctional folds are invaded and destroyed by macrophages (Fig. 8.10). The nerve terminals are never attacked by the phagocytic cells. The affected muscle fibers are denervated because they are separated from the nerve terminal by the invading macrophages. A proportion of muscle fibers undergo segmental necrosis which centers around the abnormal end-plate region.

After day 11, the inflammatory cells vanish from muscle, the nerve terminals return to the hightly simplified postsynaptic regions, and the junctional folds regenerate. During this period the animals recover clinically.

8.9.2 Chronic EAMG

During the chronic phase of EAMG which begins about day 30, junctional folds again degenerate (Fig. 8.11) but the abnormal postsynaptic regions do not split away from the underlying muscle fiber and there is no macrophage invasion. During chronic EAMG, the end-plates undergo cycles of degeneration and regeneration and immature junctions with poorly differentiated postsynaptic regions and nerve sprouts near the end-plates can be observed. Morphometric analysis shows a significant decrease in the length of the postsynaptic membrane. Minor morphometric alterations also occur in the nerve terminal, most important of which is an increased numerical density of the synaptic vesicles. This may represent a presynaptic adaptation to the postsynaptic impairment of neuromuscular transmission.

Fig. 8.11 Chronic EAMG. Numerous junctional folds are abnormally short due to shedding of their terminal expansions into the synaptic space. Widened synaptic space contains debris (x) derived from the junctional folds. × 17 000. Reproduced by permission from Engel *et al*. (1976a).

The ultrastructural evidence clearly shows that postsynaptic membrane is the primary target of the autoimmune attack in acute and chronic EAMG. End-plates of rabbits immunized with multiple doses of AChR show changes similar to those found in rats with chronic EAMG (Thornell *et al.*, 1976). More important is the fact that morphologically chronic EAMG resembles human MG (as will be shown later). However, the macrophage response of acute EAMG is observed but rarely in human MG. We have seen this phenomenon at a few end-plates in a patient who had severe MG and also a thymoma (Engel *et al.*, 1979).

8.9.3 Immunoelectron microscopic study of EAMG

Peroxidase-labeled α-bungarotoxin can be used to study the abundance and distribution of AChR at the end-plate at the ultrastructural level. This approach demonstrates a marked decrease of AChR at the end-plate in chronic EAMG (Fig. 8.12) (Engel *et al.*, 1977a).

Immunoelectron microscopic localization of IgG at the EAMG end-plate was accomplished with rabbit anti-(rat IgG) followed by treatment with peroxidase-labeled staphylococcal protein A (Sahashi *et al.*, 1978). IgG deposits are found on terminal expansion of junctional folds, where AChR is known to be located, over patches of the simplified postsynaptic membrane

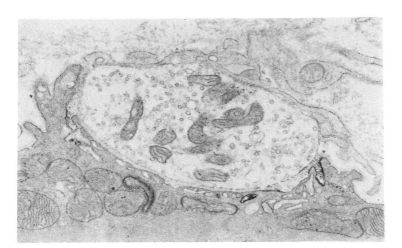

Fig. 8.12 Ultrastructural localization of AChR in chronic EAMG, rat forelimb muscle. Only short segments of the highly simplified postsynaptic membrane react for AChR. Presynaptic membrane is unstained except where it faces reactive segments of postsynaptic membrane. × 14 900. Reproduced by permission from Engel *et al.* (1977a).

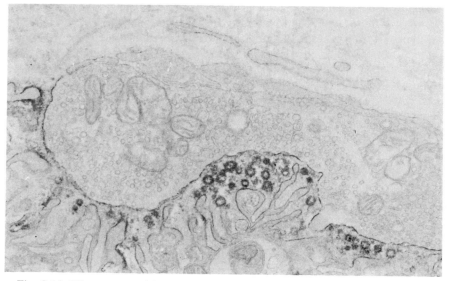

Fig. 8.13 Ultrastructural localization of IgG in chronic EAMG, rat forelimb muscle. IgG deposits are present on globular fragments shed into the synaptic space and on short segments of junctional folds. Note reciprocal staining of presynaptic membrane. × 24 900. Reproduced by permission from Sahashi *et al.* (1978).

where junctional folds are no longer present, and on membrane fragments shed into the synaptic space by disintegrating junctional folds (Fig. 8.13).

The third component of complement (C3) was also localized at the EAMG end-plate with peroxidase-labeled rabbit anti-(rat C3) (Sahashi *et al.*, 1978). The ultrastructural localization of C3 is virtually identical with that of IgG (Fig. 8.14). This indicates that anti-AChR antibodies readily fix complement *in situ*, that the complement reaction proceeds at least through its initial assembly phase, and that complement may be responsible for lysis of the junctional folds. The presence of C3, IgG and AChR on membranous material shed into the synaptic space supports the idea that complement attacks the postsynaptic membrane. Rigorous proof of this requires localization of C9 at the end-plate. This was not feasible in EAMG for lack of antibody to rat C9, but was accomplished in human MG, as will be shown subsequently.

8.9.4 Passively transferred EAMG

Passive transfer of EAMG from chronically affected donors to normal recipients recapitulates features of acute EAMG. Once again, we traced AChR (Fig. 8.15), IgG and C3 (Fig. 8.16) with specific cytochemical probes and

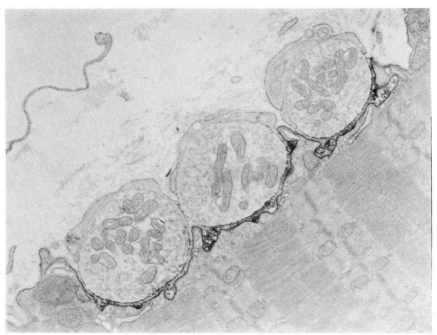

Fig. 8.14 Ultrastructural localization of C3 in chronic EAMG, rat forelimb muscle. Reaction for C3 occurs on segments of the highly simplified postsynaptic membrane and on debris in synaptic space. × 13 900. Reproduced by permission from Sahashi *et al*. (1978).

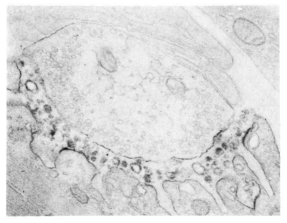

Fig. 8.15 Ultrastructural localization of AChR, 24 hours after passive transfer of EAMG, rat forelimb muscle. Globular material shed from tips of junctional folds reacts vividly for AChR. Short segments of underlying junctional folds also react. × 26 400. Reproduced by permission from Engel *et al*. (1979a).

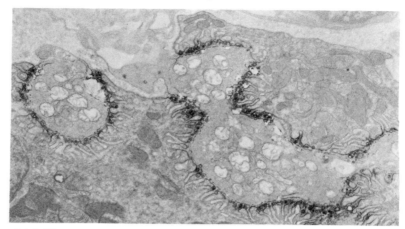

Fig. 8.16 Ultrastructural localization of C3, 48 hours after passive transfer of EAMG, rat forelimb muscle. There is intense reaction for C3 on globular fragments of junctional folds that had been shed into the synaptic space. This endplate region did not split away from its parent fiber and was not invaded by macrophages – events commonly seen by 48 hours in passively transferred EAMG. Swollen mitochondria in nerve terminal represent fixation artifact. × 10 000. Copyright 1980, Mayo Clinic.

followed the ultrastructural change by morphometric analysis (Engel *et al.*, 1979). Beginning on day one and peaking on days two and three after transfer, there is focal degeneration of junctional folds followed rapidly by invasion of the end-plate by macrophages. IgG and C3 are detected on terminal expansions of the junctional folds as early as six hours after transfer. By 24 hours, segments of folds rich in AChR and coated with IgG (Fig. 8.15) and C3 (Fig. 8.16) are shed into the synaptic space. By day two, many sensitized postsynaptic regions are destroyed by macrophages. By day five, the cellular reaction subsides and nerve terminals return to the muscle fiber. On day 10, most end-plates are highly simplified and show AChR deficiency but the animals are clinically recovered. The findings on day 10 show that there is a high safety margin of neuromuscular transmission in the rat and that clinical evaluation alone may not reveal signs of the autoimmune process at the end-plate. Fifty-four days after passive transfer, the postsynaptic regions are still smaller than normal and there is a slight decrease in postsynaptic AChR. Throughout the study, the mepp amplitude varied directly with morphometric estimates of the abundance of the postsynaptic membrane reacting for AChR (Engel *et al.*, 1979). In passively transferred EAMG, the AChR deficiency is due to lysis of the postsynaptic membrane and destruction of opsonized postsynaptic regions by macrophages. The temporal sequence of events in this syndrome is such that neither a blocking of AChR function by antibody nor antibody-dependent modulation of AChR can be important in the pathogenesis.

8.10 CLINICAL FEATURES OF MG

Clinically, MG is characterized by excessive fatigability and weakness of voluntary muscles. Weakness is worsened by exercise and partially relieved by rest. Drooping eyelids, double vision, lack of facial expression and difficult swallowing are often observed. In severe generalized cases, limb muscles and the muscles of respiration may be affected. Neonatal MG is a transient form of weakness observed in 12 per cent of infants born to myasthenic mothers (Namba *et al.*, 1970). It is caused transfer of maternal anti-AChR antibody (Keesey and Lindstrom, 1977; Keesey *et al.*, 1977). Congenital MG represents a heterogenous group in which symptoms similar to MG result from different mutations affecting neuromuscular transmission, rather than from an immune response to AChR (Engel *et al.*, 1977c, 1979b; Vincent and Newsome-Davis, 1979; Hart *et al.*, 1979).

MG is a rare disease affecting only about one in 20 000 of the population (Osserman and Genkins, 1971). Younger females and older males are most frequently affected. This may indicate an effect of androgens and estrogens on susceptibility of the immune system to autoimmune response (cf. Talal, 1977). In Caucasians there is an increased frequency of HLA-B8 and HLA-DR w 3 (Pirskanen, 1976; Naeim *et al.*, 1978). Similar disequilibria are also noted in other autoimmune diseases. This may indicate a genetic predisposition to autoimmune response in some patients, as does the observation of antimuscle striation antibodies in MG patients with thymoma (Oosterhuis *et al.*, 1976) and an increased incidence of other autoimmune diseases in patients with MG (Simpson, 1960). Ten to fifteen per cent of MG patients have thymomas, and in 72 per cent thymic germinal centers suggestive of local antigenic stimulation or antibody production are observed (Papatestas *et al.*, 1976). Thymectomy is of some therapeutic benefit in MG (Buckingham *et al.*, 1976). Muscle cells can be cultured from thymus tissue (Wekerle *et al.*, Kao and Drachman, 1976) and very small amounts of AChR can be found in thymus (Aharonov *et al.*, 1975b). These results may indicate that in MG the thymus provides a source of antigenic stimulation or peculiarly autoresponsive cells. Patients with rheumatoid arthritis treated with D-penicillamine may occasionally develop mild MG (Bucknall *et al.*, 1975). This is associated with the development of serum anti-AChR antibodies (Vincent *et al.*, 1978; Russell and Lindstrom, 1978). Antibodies to muscle striations may also be induced in these patients (Peers *et al.*, 1977). The weakness and the anti-AChR antibodies concomitantly diminish after cessation of D-penicillamine. Penicillamine can trigger several autoimmune diseases (Fernandes *et al.*, 1977; Schrader *et al.*, 1972; Engel *et al.*, 1973). In the case of MG, it is unknown whether penicillamine modifies AChR to make it immunogenic, or effects regulatory or responder immune cells directly to make the latter hyper-responsive. Antibodies to polyadenylic acid are found in half of the MG patients tested (Fishbach *et al.*, 1981). Whether this reflects a defect in immune regulation, is secondary to a lytic viral infection, or is caused by some other mechanism is unknown. Despite all these tantalizing clues, the cause (or causes) of MG is (are) not known. Determining the cause of MG may be a very difficult problem. Although EAMG is a very valuable model for pathological mechanisms at the end-plate, it seems much less valuable for studying what actually causes the immune response to AChR in patients with MG.

As symptomatic therapy, MG patients are very frequently given drugs which inhibit acetylcholinesterase in order to attempt to compensate for the decrease in acetylcholine sensitivity of the muscle by increasing the concentration and lifetime in the synaptic space of the released transmitter. Acute overdose with acetylcholinesterase inhibitors can impair transmission, probably by accumulation of desensitized AChR. Chronic and massive overdoses of normal animals results in disruption of the junctional folds which superficially resembles that in MG (Salpeter *et al.*, 1979).

In order to suppress the aberrant immune response to AChR in MG, patients are often treated with some combination of thymectomy (Buckingham *et al.*, 1976; Seybold *et al.*, 1978; Vincent *et al.*, 1979), corticosteroid therapy (Sanders *et al.*, 1979; Seybold and Drachman, 1974; Seybold and Lindstrom, 1979), cytotoxic drugs (Hertel *et al.*, 1979; Reuther *et al.*, 1979) and lymph drainage (Lefvert and Matell, 1979) or plasmapheresis (Pinching *et al.*, 1976; Newsom-Davis *et al.*, 1978, 1979; Dau *et al.*, 1977, 1979) in combination with immunosuppressive drugs. These approaches vary in their efficacy, and each has its passionate proponents and vehement detractors. None of the therapies is specifically directed at what instigates or propagates the immune response to AChR, none provides a cure, and all have real or potential unfortunate side effects. Most interesting for our purposes is that simply exchanging large amounts of plasma for a substitute obviously reduces serum anti-AChR concentration, and is associated with a beneficial clinical response (Newsom-Davis *et al.*, 1978, 1979; Dau *et al.*, 1977, 1979). This observation, along with observation of passively transferred anti-AChR antibodies in neonatal MG (Kessey *et al.*, 1977), and passive transfer of MG to mice with anti-AChR IgG (Tokya *et al.*, 1975, 1977, 1978), helps to demonstrate the importance of the antibody-mediated immune response in MG.

8.11 PATHOLOGICAL MECHANISMS IN MG

Antibodies reactive with AChR from human muscle are detectable at levels averaging about 100-fold the normal background in about 90 per cent of patients thought to have MG (Lindstrom, 1977; Lindstrom *et al.*, 1976a). The human muscle used in the assays is usually from amputated limbs of diabetic patients or other cases with associated muscle denervation. This muscle may contain extrajunctional AChR, which is more antigenic for MG sera than junctional AChR (Weinberg and Hall, 1979). Several variations on the basic method for assaying antibodies to AChR (Patrick *et al.*, 1973) have been reported for assaying antibodies in MG sera (Dwyer *et al.*, 1979; Mittag *et al.*, 1976; Monnier and Fulpius, 1977; Sondag-Tschroots *et al.*, 1979; Appel *et al.*, 1975; Aharonov *et al.*, 1975c). If AChR from other species is used as antigen, detection of antibodies is less effective than with human AChR, because the anti-AChR in MG patients, like that in animals with EAMG, is very species specific (Lindstrom *et al.*, 1978b; Marengo *et al.*, 1979). As in EAMG, most anti-AChR antibodies from MG patients are directed at determinants other than the acetylcholine binding site (Lindstrom *et al.*, 1970a). The assay used selectively detects high affinity ($K_D \sim 10^{-9}$ M) antibodies, thus antibodies would not be detected in a patient with a 1×10^{-7} M concentration of antibodies with a K_D of 1×10^{-7} M, yet half of his AChRs would be saturated with antibody. Absolute concentration of anti-AChR in serum varies widely and does not correlate closely with disease severity, though patients with only ocular symptoms have significantly lower titers (Lindstrom *et al.*, 1979a). In individuals treated by plasmapheresis (Lefvert and Matell, 1979; Newsom-Davis *et al.*, 1978, 1979; Dau *et al.*, 1977, 1979), immuno-suppressive drugs (Hertel *et al.*, 1979; Lefvert and Matell, 1979), or patients suffering from neonatal MG (Keesey and Lindstrom, 1977; Keesey *et al.*, 1977; Halcao *et al.*, 1977) or penicillamine induced MG (Vincent *et al.*, 1978b; Russell and Lindstrom, 1978), a clear correlation between changes in antibody titer and clinical state is observed. Independent of the therapy used, clinical improvement is associated with a sustained 50–75 per cent decrease in antibody titer (Seybold and Lindstrom, 1981). Synthesis of antibodies to AChR by thymic lymphocytes can be detected *in vitro* (Vincent *et al.*, 1978a, 1979).

Peripheral lymphocytes of MG patients can be stimulated to incorporate thymidine by electric organ AChR (Abramsky *et al.*, 1975a,b; Richman *et al.*, 1976, 1979; Conti-Tronconi *et al.*, 1977, 1979; Morgutti *et al.*, 1979). Unfortunately, due to the limited cross-reaction between electric organ and human AChR, only about a third of patients show a response above background, the average is only twofold above background, and there are false positives. Among the few patients showing a response to electric organ AChR, there is no clear correlation between absolute response and clinical state, but responses do diminish after thymectomy or immunosuppressive therapy (Abramsky *et al.*, 1975b; Morgutti *et al.*, 1979).

In MG patients, stimulation of nerves with extracellular electrodes at 2–3 Hz results in a characteristic decrease in the number of muscle fibers successfully completing neuromuscular transmission with each stimulus (Özdemir and Young, 1976). Micro-electrode studies of isolated intercostal muscle fibers show that the number of quanta released by nerve impulse is normal, but that the average amplitude of the mepp is reduced to about one third of normal (Elmquist *et al.*, 1964; Lambert and Elmquist, 1971). Ionotophoresis of acetylcholine shows that the sensitivity of the postsynaptic membrane is decreased (Albquerque *et al.*, 1976). In man, normally about 60 quanta are released by 1 Hz stimulation of the nerve (Lambert and Elmquist, 1971), each containing about 10^4 molecules of acetylcholine (Kuffler and Yoshikami, 1975). Normally a quantum of acetylcholine activates 1500 AChR, whereas in the average MG patient it activates 500 (Cull-Candy *et al.*, 1979). The function of these AChR is impaired little (Bevan *et al.*, 1978) if at all (Cull-Candy *et al.*, 1979) by the antibodies which are bound in MG patients.

In the average MG patient, the AChR content of intercostal muscle is about 36 per cent of normal (Lindstrom and Lambert, 1978) or about 2×10^6 AChR molecules per end-plate (Fambrough *et al.*, 1973). An average of 51 per cent of the AChR which remain have antibodies bound (Lindstrom and Lambert, 1978). The acetylcholine sensitivity is directly proportional to the amount of AChR remaining (Engel *et al.*, 1977b; Lindstrom and Lambert, 1978).

As will be detailed in the next section, the morphological and immunoelectron microscopic features of the end-plate in human MG resemble those in the rat forelimb muscle in chronic EAMG.

In MG, as in EAMG (Heinemann *et al.*, 1977; Merlie *et al.*, 1979b; Lindstrom and Einarson, 1979), antigenic modulation probably causes much of the observed loss of AChR. Sera from MG patients increase the rate of AChR degradation in cultured rat (Kao and Drachman, 1977; Drachman *et al.*, 1978a,b; Appel *et al.*, 1977) and human (Bevan *et al.*, 1977) muscle cells by crosslinking AChR (Drachman *et al.*, 1978b) and thereby triggering its internalization and lysosomal degradation.

The pathological mechanisms by which AChR content and acetylcholine sensitivity is reduced in MG appear to be fundamentally the same in MG and chronic EAMG. The most critical difference between EAMG and MG appears to be that the autoimmune response to AChR in EAMG results from the small but significant cross-reaction of antibodies to electric organ AChR with muscle AChR, and the autoimmune response parallels the course of the response to the foreign immunogen, whereas the MG antibodies appear to be directed primarily at human AChR and arise and are sustained by an unknown endogenous process. Another critical difference is that EAMG can be studied in young, female, inbred rats, whereas MG occurs in patients differing in age, sex, genetic background, and states of disease history.

8.12 MORPHOLOGICAL FEATURES AND IMMUNO-PATHOLOGY OF THE NEUROMUSCULAR JUNCTION IN MG

8.12.1 Morphometric reconstruction of the MG end-plate

Earlier ultrastructural studies of the MG end-plate described a variety of abnormalities (reviewed by Engel and Santa, 1971), but none were quantitative and none assessed adequately possible variations of fine structure at the normal end-plate. Between 1968 and 1970, Engel and Santa (1971) developed a system for the morphometric analysis of the fine structure of the motor end-plate and applied this method to specimens of external intercostal muscle from control subjects, and patients with MG and with the Lambert–Eaton myasthenic syndrome. (In the latter disorder mepp is of normal amplitude but neuromuscular transmission fails due to an abnormally low quantum content of the epp.) This study was done at a time when the prevailing notion was that the small amplitude of the mepp in MG was due to a presynaptic defect in ACh synthesis or packaging. If this were the case, one might expect changes in the dimensions of the synaptic vesicles (Jones and Kwanbunbumpen, 1970). Contrary to what was expected, the dimensions, and also the numerical density, of the synaptic vesicles was entirely normal. The nerve terminals were smaller than normal but their structural integrity was maintained. By contrast, the postsynaptic regions showed degenerative changes with widening of the primary and secondary synaptic clefts and accumulation of debris in the synaptic space (Fig. 8.17) (Engel and Santa, 1971). Morphometric analysis revealed a significant decrease in the length of

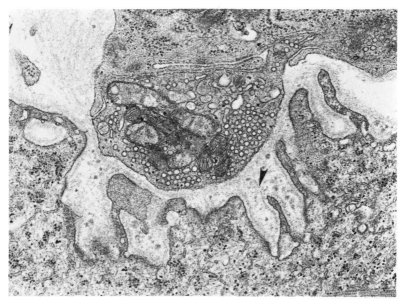

Fig. 8.17 End-plate region from external intercostal muscle of patient with untreated MG. The synaptic space is widened due to degeneration of the junctional folds. Degenerate remnants of folds and traces of basement membrane (arrowhead) which invested pre-existing folds persist. × 31 900. Reproduced by permission from Engel *et al.* (1976a).

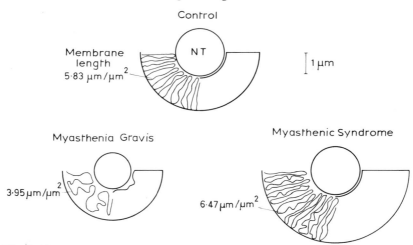

Fig. 8.18 Morphometric reconstruction of the motor end-plate in non-weak controls, MG and the Lambert–Eaton myasthenic syndrome. Schwann cell and junctional sarcoplasm are not shown. Reproduced by permission from Engel and Santa (1971).

the postysynaptic membrane. Finally, there were nerve-terminal sprouts, highly degenerated postsynaptic regions denuded of their nerve terminals, and immature junctions with poorly differentiated postsynaptic regions. All these suggested cycles of degeneration and regeneration occurring at the MG end-plate. In the Lambert–Eaton syndrome the postsynaptic region was somewhat larger than normal, the postsynaptic membrane was increased in length, and there were no presynaptic or postsynaptic degenerative changes. Fig. 8.18 shows morphometric reconstruction of the motor end-plate in control subjects, in MG and in the Lambert–Eaton syndrome. In the light of current knowledge, the postsynaptic pathology of MG is a direct consequence of the autoimmune attack on the postsynaptic AChR.

8.12.2 Ultrastructural localization and quantification of AChR in MG

Radiochemical and light-microscopic radioautographic evidence of AChR deficiency at the MG end-plate has been obtained by Fambrough *et al*. (1973). The ultrastructural localization of AChR at the MG end-plate is also of interest because this allows (1) direct comparison of the abundance and distribution of AChR at the normal vs MG end-plate; (2) correlation of changes in end-plate fine structure with alterations in receptor distribution; and (3) correlations of morphometric estimates of AChR with the mepp amplitude and with the clinical state. AChR was localized with peroxidase-labeled α-bungarotoxin applied to fresh, well-oxygenated, intact (from origin to insertion) muscle strips, well rinsed before and after exposure to the reagent, followed by glutaraldehyde fixation. In this way, sensitivity of the reaction and preservation of fine structure were both optimized. External inter-

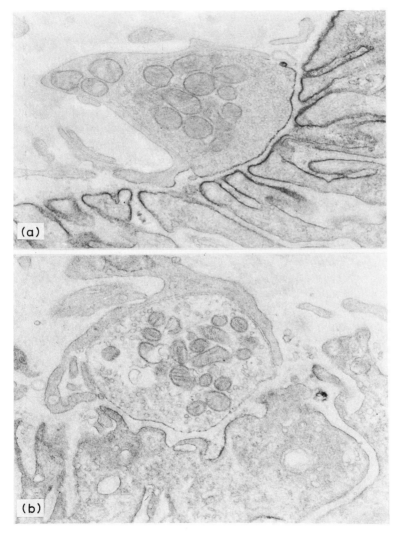

Fig. 8.19 AChR localization in control subject (a) and in patient with moderately severe, generalized MG (b). There is marked deficiency of postsynaptic AChR in (b). Presynaptic staining, seen in (a), is essentially absent in (b). × 22 300. Reproduced by permission from Engel *et al.* (1977a).

costal muscle strips were studied from 10 patients with MG (196 end-plates). Seven patients with the Lambert–Eaton myasthenic syndrome (175 end-plates) and four non-weak subjects (175 end-plates) serves as controls (Engel *et al.*, 1979a, 1979).

In MG, the ultrastructural localization of AChR is abnormal in that some junctional folds show no or only trace reaction for AChR; and at some end-plate regions only short segments of the postsynaptic membrane react for AChR (Fig. 8.19). Usually, but not always, the simplest and most degenerated postsynaptic regions show the greatest decrease in toxin binding. There is also presynaptic staining for AChR but, as a rule, this is reduced or absent when vis-a-vis junctional folds do not react for AChR (Fig. 8.19). This suggests that the presynaptic staining is largely artifactual caused by diffusion of reaction product from postsynaptic sites.

In the Lambert–Eaton syndrome the abundance and distribution of AChR at the end-plate is normal. The observations can be quantified by measuring the length of the postsynaptic membrane reacting for AChR at a given end-plate region and normalizing this for the length of the primary synaptic cleft. The value thus obtained is referred to as the AChR index. In our series of MG patients the mean AChR index ranges from 0.27 to 1.62 (group mean ±SEM: 0.85 ± 0.14). In the non-weak controls the corresponding values range from 3.06 ± 3.29 (3.21 ± 0.05); and in the Lambert–Eaton syndrome from 2.56 to

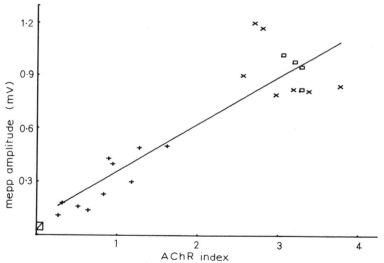

Fig. 8.20 Regression of the mean mepp amplitude on the mean AChR index in 10 cases of MG (+), 7 cases of the Lambert–Eaton myasthenic syndrome (x) and 4 controls (o). The regression line is linear, with *P* ≪ 0.001. Reproduced by permission from Engel *et al.* (1979).

3.77 (3.05 ± 0.16). In collaboration with E.H. Lambert, we compared the mean AChR indices obtained in each subject with the corresponding mean mepp amplitudes (Engel *et al.*, 1979a, 1979). The comparison demonstrated a highly significant linear correlation between the two variables ($r = 0.899$; $P \ll 0.001$) (Fig. 8.20).

These findings in MG give unambiguous ultrastructural evidence for a deficiency of postsynaptic AChR. They also indicate that the AChR deficiency determines the small mepp amplitude and hence the defect of neuromuscular transmission. A linear relationship between the AChR content of muscle and the mepp amplitude has been demonstrated in further studies in which the AChR content of muscle homogenates was radiochemically assayed (Lindstrom and Lambert, 1977; Ito *et al.*, 1978).

8.12.3 Ultrastructural localization of IgG at the MG end-plate

Although evidence for an autoimmune pathogenesis of acquired MG had been accumulating since 1973, no immunoglobulin was demonstrated at the MG end-plate before 1977. We succeeded in localizing IgG at the MG end-plate with peroxidase-labeled staphylococcal protein A (Engel *et al.*, 1977d). Protein A binds bivalently to Fc portions of human IgG subclasses 1, 2 and 4 and has an increased affinity for IgG molecules attached to antigenic sites (Kessler, 1976). The latter property enhances the specificity and sensitivity of the reagent and optimizes the signal (reaction at the end-plate) to noise (background staining) ratio of the cytochemical procedure.

As in the preceding study, the immunoreagent was applied to fresh, intact muscle strips, well rinsed before and after the application of the immunoreagent. In less severe cases of MG, in which the postsynaptic folds were relatively well preserved, IgG was found over terminal expansions of the junctional folds where AChR is known to be located (Fig. 8.21a) or over short segments of the postsynaptic membrane (Fig. 8.22). In more severe cases, in which junctional folds were less well preserved, the postsynaptic membrane bound less IgG, but IgG was readily detected on degenerated remnants of the junctional folds in the synaptic space (Fig. 8.21b). Morphometric estimates of the abundance of the postsynaptic membrane binding IgG were proportionate to the mepp amplitude (Engel *et al.*, 1977d), but an exception to this relationship was noted in one of ten cases. This was a young patient who, despite having moderately severe disease, had well-preserved junctional folds which bound abundant IgG.

These findings indicate that (1) the amount of IgG bound to the postsynaptic membrane is proportionate to the amount of AChR which remains in the membrane; (2) IgG-coated segments of junctional folds are shed into the synaptic space; (3) in most cases of MG the transmission defect is determined by the loss of AChR rather than by the fact that IgG is bound to AChR; (4) in

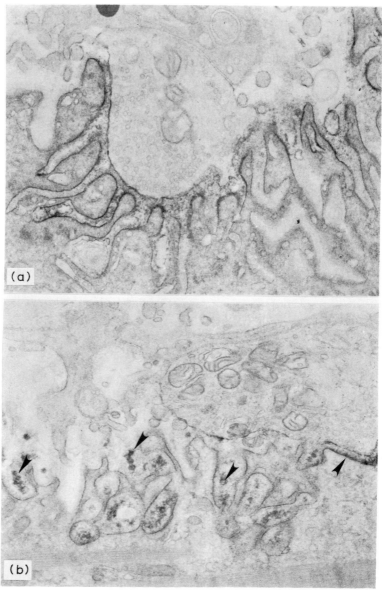

Fig. 8.21 Ultrastructural localization of IgG in mild (a) and more severe (b) MG. In (a), abundant IgG is found on terminal expansions of well-preserved junctional folds. In (b), IgG is found mostly on degenerated material accumulating in the synaptic space and on short segments of the postsynaptic membrane (arrows). (a) and (b). × 25 600. Reproduced by permission from Engel *et al*. (1977d).

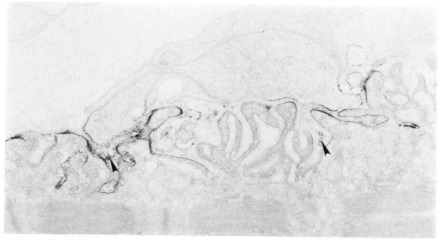

Fig. 8.22 IgG localization in MG. Reaction is found on short segments of the junctional folds and on debris in the synaptic space (arrowheads). Note reciprocal staining of presynaptic membrane where it faces reactive segments of postsynaptic membrane. This suggests that presynaptic staining represents diffusion artifact of cytochemical procedure. × 27 300. Reproduced by permission from Engel *et al*. (1977d).

most cases of MG, antibodies to AChR do not significantly block the response of AChR to ACh, but in occasional patients antibodies may have a significant blocking effect.

8.12.4 Participation of the complement system in the autoimmune attack on the postsynaptic membrane

We sought evidence for antibody-dependent complement mediated injury to the junctional folds by localizing complement components at the MG end-plate (Engel *et al*., 1979d, 1979; Sahashi *et al*., 1980). The third (C3) and ninth (C9) complement components were localized with monospecific antibodies raised in rabbits. The ultrastructural localization of C3 was identical to that of IgG. Reaction product was detected on segments of the postsynaptic membrane, on debris in the synaptic space and on disintegrating junctional folds (Engel *et al*., 1977d). The fact that the distribution of C3 and IgG at the end-plate are highly similar is consistent with the assumption that anti-AChR antibodies can fix complement and that the assembly phase of the complement reaction sequence (via C1, C4, C2 and C3) has gone to completion. However, the presence of C3 on the junctional folds does not in itself establish that the attack (or lytic) phase of the complement reaction has become activated (Müller-Eberhard, 1975).

The occurrence of complement-mediated membrane injury depends on the presence of the target membrane of C5b, C6, C7, C8 and C9, the constituents of the membrane attack complex. The complex, now known to be a dimer of C5b-9, induces focal and irreversible membrane lesions and C9 is an absolute requirement for this (Müller-Eberhard, 1975; Biesecker *et al.*, 1979). Therefore, to obtain unambiguous evidence for complement-mediated injury of the

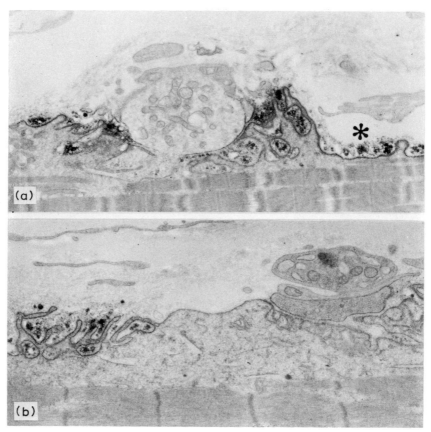

Fig. 8.23 Ultrastructural localization of C9, the lytic component of the complement sequence, at MG end-plates. In (a), reaction for C9 appears on debris in the synaptic space, over short segments of the junctional folds and on material to the right of the folds, on the surface of the muscle fiber, positioned between layers of basal lamina (asterisk). In (b), end-plate region on right shows only trace reaction for C9. Postsynaptic region at left is denuded of its nerve terminal and consists of debris and degenerating folds. Strong reaction for C9 occurs on debris near folds. (a) × 11 000; (b) × 13 100. Reproduced by permission from Sahashi *et al.* (1980).

postsynaptic membrane, we decided to study the localization of C9 at the MG end-plate (Sahashi *et al.*, 1980).

At end-plates where the architecture of the postsynaptic region was well preserved, C9 was localized on short segments of the junctional folds and on sparse debris in the synaptic space. At end-plates which showed degenerative changes, there was intense reaction for C9 on degenerate material in the synaptic space and this was most marked where the junctional folds were the most abnormal (Fig. 8.23). Reaction for C9 also occurred near existing end-plates, on debris positioned between layers of basal lamina. Finally, intense reaction for C9 was found at highly degenerate postsynaptic regions denuded of their nerve terminal (Fig. 8.23). Here the reaction was on debris which often was still arranged in the shape of the pre-existing junctional fold and was still surrounded by traces of basal lamina. On some muscle fibers discrete regions reacting for C9 were widely separated. Some of these were postsynaptic regions covered by nerve terminal; some consisted of degenerate folds and debris not covered by nerve terminal; and some consisted only of debris sandwiched between layers of basal lamina. In contrast with the localization of IgG and C3, no definite relationships could be established between the abundance of the reaction for C9 (most of which was associated with debris) and the severity of the disease or the amplitude of the mepp.

These findings indicate that (1) the presence of IgG or C3 on the postsynaptic membrane does not in and of itself injure the junctional folds; (2) the complement reaction sequence, subject to constraints of time and space, does not go to completion at all sites that had bound C3; (3) once formed, the C5b-9 dimer is more stable than the membrane which it attacks; (4) segments of the junctional folds attacked by the C5b-9 complex are shed into the synaptic space where they further disintegrate; (5) remodeling of the end-plate and progressive separation of end-plate regions occur when the nerve terminal moves away from junctional folds destroyed by complement to an adjacent site where a new end-plate region develops; (6) complement mediated lysis of the junctional folds, like modulation (Heinemann *et al.*, 1977; Merlie *et al.*, 1979b; Lindstrom and Einarson, 1979), contributes to the receptor deficiency at the MG end-plate.

REFERENCES

Abramsky, O., Aharonov, A., Webb, C. and Fuchs, S. (1975a), *Clin. Exp. Immunol.*, **19**, 11–16.

Abramsky, O., Aharonov, A., Teitelbaum, D. and Fuchs, S. (1975b), *Arch. Neurology*, **32**, 684–687.

Aharonov, A., Kalderon, N., Silman, I. and Fuchs, S. (1975a), *Immunochem.*, **12**, 765–771.

Aharonov, A., Tarrab-Hazdai, R., Abramsky, O. and Fuchs, S. (1975b), *Proc. Natl. Acad. Sci. USA,* **72**, 1456–1459.

Aharonov, A., Abramsky, O., Tarrab-Hazdai, R. and Fuchs, S. (1975c), *Lancet,* **1**, 340–342.

Albquerque, E.X., Rash, J.E., Meyer, R.F. and Satterfield, J.R. (1976), *Exp. Neurol.,* **51**, 536–563.

Allen, T. and Potter, L.T. (1977), *Tissue & Cell,* **9**, 609–622.

Anderson, C.R. and Stevens, C.F. (1973), *J. Physiol.,* **235**, 655–691.

Anholt, R., Lindstrom, J. and Montal, M. (1980), *Eur. J. Biochem.,* **109**, 481–487.

Appel, S.H., Almon, R.R. and Levy, N. (1975), *New Engl. J. Med.,* **293**, 760–761.

Appel, S.H., Anwyl, R., McAdams, M.W. and Elias, S. (1977), *Proc. Natl. Acad. Sci. USA,* **74**, 2130–2134.

Bennett, M.V.L. (1970), *Annu. Rev. Physiol.,* **32**, 471–528.

Bennett, M.R. and Florin, T. (1974), *J. Physiol.,* **238**, 93–107.

Berman, P.W. and Patrick, J. (1980), *J. Exp. Med.,* **151**, 204–223.

Berti, F., Clementi, F., Conti-Tronconi, B. and Folco, G. (1976), *Br. J. Pharmacol.,* **57**, 17–22.

Bevan, S., Heinemann, S., Lennon, V.A. and Lindstrom, J. (1976), *Nature,* **260**, 438–439.

Bevan, S., Kullberg, R.W. and Heinemann, S. (1977), *Nature,* **267**, 264–265.

Bevan, S., Kulberg, R.W. and Rice, J. (1978), *Nature,* **273**, 469–470.

Bickerstaff, E.R. and Woolf, A.L. (1960), *Brain,* **83**, 10–23.

Biesecker, G., Podack, E.R., Halverson, C.A. and Müller-Eberhard, H.J. (1979), *J. Exp. Med.,* **149**, 448–458.

Buckingham, J., Howard, F.M., Bernatz, P., Payne, W.S., Harrison, E.G., Obrien, B.C. and Weiland, L.H. (1976), *Ann. Surg.,* **184**, 453–458.

Bucknall, R.C., Dixon, A.J., Glieb, E.N., Woodland, J. and Zutshi, D.W. (1975), *Brit. Med. J.,* **1**, 600–602.

Cartaud, J., Benedetti, E., Sobel, A. and Changeux, J.P. (1978), *J. Cell Sci.,* **29**, 313–337.

Chang, H.W. and Bock, E. (1977), *Biochemistry,* **16**, 4513–4520.

Changeux, J.P., Heidmann, T., Popot, J. and Sobel, A. (1979), *FEBS Lett.,* **105**, 181–187.

Christadoss, P., Lennon, V. and David, C. (1979a), *J. Immunol.,* **123**, 2540–2543.

Christadoss, P., Lennon, V., Lambert, E. and David, C. (1979b), in *T and B Lymphocytes: Recognition and Function* (Bach, F., Bonavida, B. and Villeta, E. eds), pp. 249–256.

Clark, A.W., Hurlbut, W.P. and Mauro, A. (1972), *J. Cell Biol.,* **52**, 1–14.

Claudio T. and Raftery, M.A. (1977), *Arch. Biochem. Biophys.,* **181**, 484–489.

Conti-Tronconi, B., Padova, F., Morgutti, M., Missiroli, A. and Frattola, L. (1977), *Neurology,* **29**, 496–501.

Conti-Tronconi, B., Padova, F., Morgutti, M., Missiroli, A. and Frattola, L. (1977), *J. Neuropathol. Exp. Neurol.,* **36**, 157–168.

Cull-Candy, S., Miledi, R. and Trautmann, A. (1979), *J. Physiol.,* **287**, 247–265.

Cull-Candy, S., Miledi, R., Trautman, A. and Uchitel, O. (1980), *J. Physiol.,* **299**, 621–638.

Damle, V. and Karlin, A. (1978), *Biochemistry,* **17**, 2039–2045.

Damle, V., McLaughlin, M. and Karlin, A. (1978), *Biochem. Biophys. Res. Commun.*, **84**, 845–851.

Dau, P.C., Lindstrom, J.M., Cassell, C.K. and Clark, E.C. (1979), in *Plasmapheresis and the Immunobiology of Myasthenia Gravis* (Dau, P. ed.), Houghton Mifflin, Boston, pp. 229–247.

Dau, P.C., Lindstrom, J.M., Cassel, C.K., Denys, E.H., Shev, E. and Spitler, L.E. (1977), *New Engl. J. Med.*, **297**, 1134–1140.

Del Castillo, J. and Katz, B. (1954), *J. Physiol.*, **124**, 560–573.

Desmedt, J. and Borenstein, S. (1976), *Ann. N.Y. Acad. Sci.*, **274**, 174–188.

Dionne, V., Steinbach, J. and Stevens, C. (1978), *J. Physiol.*, **281**, 421–444.

Drachman, D.B., Angus, C.W., Adams, R.N. and Kao, I. (1978a), *Proc. Natl. Acad. Sci. USA*, **75**, 3422–3426.

Drachman, D.B., Angus, C.W., Adams, R.N., Michelson, J.D. and Hoffman, G.J. (1978b), *New Engl. J. Med.*, **298**, 1116–1122.

Dwyer, D.S., Bradley, R.J., Oh, S.J. and Kemp, G.E. (1979), *Clin. Exp. Immunol.*, **37**, 448–451.

Ellisman, M.H., Rash, J.E., Staehlin, L.A. and Porter, K.R. (1976), *J. Cell Biol.*, **68**, 752–774.

Elmquist, D., Hoffman, W.W., Kugelberg, J. and Quastel, D.M.J. (1964), *J. Physiol.*, **174**, 417–434.

Elmquist, D. and Quastel, D.M.J. (1965), *J. Physiol.*, **178**, 505–529.

Engel, A.G., Lindstrom, J., Lambert, E.H. and Lennon, V.A. (1977a), *Neurology*, **27**, 307–315.

Engel, A.G., Lambert, E.H. and Howard, F.M. (1977d), *Mayo Clin. Proc.*, **52**, 267–280.

Engel, A.G., Lambert, E.H., Mulder, D.W., Torres, C.F., Sahashi, K., Bertorini, T.E. and Whitaker, J.N. (1979b), *Ann. Neurol.*, **6**, 146.

Engel, A.G., Lambert, E.H. and Gomez, M.R. (1977c), *Ann. Neurol.*, **1**, 315–330.

Engel, A.G., Lambert, E.H. and Santa, T. (1973), *Neurology*, **23**, 1273–1281.

Engel, A.G., Sahashi, K., Lambert, E. and Howard, F. (1979), *Excerpta Medica ICS*, **455**, 111–112.

Engel, A.G., Sakakibara, H., Sahashi, K., Lindstrom, J., Lambert, E. and Lennon, V. (1979a), *Neurology*, **29**, 179–188.

Engel, A.G. and Santa, T. (1971), *Ann. N.Y. Acad. Sci.*, **183**, 46–63.

Engel, A.G., Tsujihata, M. and Jerusalem, F. (1975), *Peripheral Neuropathy* (Dyck, P.J., Lambert, E.H. and Thomas, P.K., eds), W.B. Saunders, Philadelphia, pp. 1404–1415.

Engel, A.G., Tsujihata, M., Lindstrom, J. and Lennon, V. (1976a), *Ann. N.Y. Acad. Sci.*, **274**, 60–79.

Engel, A.G., Tsujihata, M., Lambert, E., Lindstrom, J. and Lennon, V. (1976b), *J. Neuropathol. Exp. Neurol.*, **35**, 569–587.

Engel, A.G., Tsujihata, M., Sakakibara, H., Lindstrom, J. and Lambert, E. (1977b), *Excerpta Medica ICS*, **404**, 132–142.

Epstein, M. and Racker, E. (1978), *J. Biol. Chem.*, **253**, 6660–6662.

Fambrough, D.M. (1979), *Physiol. Rev.*, **59**, 165–227.

Fambrough, D.M., Devreotes, P., Cord, D., Gardner, J. and Tepperman, K. (1978), *National Cancer Institute Monograph*, **48**, 277–294.

Fambrough, D.M., Drachman, D.B. and Satyamurti, S. (1973), *Science*, **182**, 293–295.

Fernandes, L., Swinson, D.R. and Hamilton, E.B. (1977), *Ann. Rheum. Dis.*, **36**, 94–95.

Fertuk, H.C. and Salpeter, M.M. (1974), *Proc. Natl. Acad. Sci. USA*, **71**, 1376–1378.

Fertuck, H.C. and Salpeter, M.M. (1976), *J. Cell Biol.*, **69**, 144–158.

Fishbach, M., Lindstrom, J., Talal, N. (1981), *Clinical and Experimental Immunology*, **43**, 73–79.

Froehner, S., Karlin, A. and Hall, Z. (1977), *Proc. Natl. Acad. Sci. USA*, **74**, 4685–4688.

Froehner, S.C. and Rafto, S. (1979), *Biochemistry*, **18**, 301–307.

Fuchs, S., Nevo, D. and Tarrab-Hazdai, R. (1976), *Nature*, **263**, 329–330.

Fulpius, B., Zurn, A., Granato, D. and Leder, R. (1976), *Ann. N.Y. Acad. Sci. USA*, **274**, 116–129.

Glavinovic, M.I. (1979), *J. Physiol.*, **290**, 481–497.

Granato, D., Fulpius, B.W. and Moody, J. (1976), *Proc. Natl. Acad. Sci. USA*, **73**, 2872–2876.

Green, D.P.L., Miledi, R. and Vincent, A. (1975), *Proc. R. Soc. Lond. (B)*, **189**, 57–68.

Hakao, K., Nishitani, H., Suzuki, M., Ohta, M. and Hayashi, K. (1977), *New Eng. J. Med.*, **297**, 169.

Hall, Z. (1973), *J. Neurobiol.*, **4**, 343–361.

Hamilton, S.L., McLaughlin, M. and Karlin, A. (1977), *Biochem. Biphys. Res. Commun.*, **79**, 692–699.

Hart, Z.H., Sahashi, K., Lambert, E.H., Engel, A.G. and Lindstrom, J. (1979), *Neurology*, **29**, 559.

Hartzell, H.C, Kuffler, S.W. and Yoshimaki, D. (1975), *J. Physiol.*, **251**, 427–463.

Heilbronn, E., Mattsson, C., Stalberg, E.J. and Hilton-Brown, P. (1975), *J. Neurol. Sci.*, **24**, 59–64.

Heinemann, S., Bevan, S., Kullberg, R., Lindstrom, J. and Rice, J. (1977), *Proc. Natl. Acad. Sci. USA*, **74**, 3090–3094.

Heinemann, S., Merlie, J. and Lindstrom, J. (1978), *Nature*, **375**, 65–68.

Hertel, G., Mertens, H.G., Reuther, P. and Ricker, K. (1979), in *Plasmapheresis and the Immunobiology of Myasthenia Gravis*, (Dau, P., ed.), Houghton Mifflin, Boston, pp. 315–328.

Heuser, J.E. and Miledi, R. (1971), *Proc. Roy. Soc. Lond. (B)*, **179**, 247–260.

Heuser, J.E. and Reese, T.S. (1973), *J. Cell Biol.*, **57**, 315–344.

Heuser, J.E., Reese, T.S., Dennis, M., Jan, Y., Jan, L. and Evans, L. (1979), *J. Cell Biol.*, **81**, 275–300.

Heuser, J.E., Reese, T.S. and Landis, D.M.D. (1974), *J. Neurocytol.*, **3**, 109–131.

Heuser, J.E. and Salpeter, S. (1979), *J. Cell Biology*, **82**, 150–173.

Huang, L.M., Catterral, W.A. and Ehrenstein, G. (1978), *J. Gen. Physiol.*, **71**, 397–410.

Hubbard, J.I. (1973), *Physiol. Rev.*, **53**, 674–723.

Huganir, R., Schell, M. and Racker, E. (1979), *FEBS Lett.*, **108**, 155–160.

Ito, Y., Miledi, R., Vincent, A. and Newsom-Davis, J. (1978), *Brain*, **101**, 345–368.

Jacob, M. and Lentz, T. (1979), *J. Cell Biol.*, **82**, 195–211.

Jones, S.F. and Kwanbunbumpen, S. (1970), *J. Physiol.*, **207**, 31–50.

Kao, I. and Drachman, D.B. (1976), *Science*, **195**, 74–75.

Kao, I. and Drachman, D.B. (1977), *Science*, **196**, 527–529.

Karlin, A., Holtzman, E., Valderrama, R., Damle, V., Han, K. and Reyes, F. (1978), *J. Cell Biol.*, **76**, 577–592.

Karlin, A., Weill, C.L., McNamee, M.G. and Valderrama, R. (1976), *Cold Spring Harbor Symp. Quant. Biol.*, **40**, 203–210.

Katz, B. (1966), *Nerve, Muscle and Synapse,* McGraw-Hill, New York.

Katz, B. and Miledi, R. (1967a), *J. Physiol.*, **189**, 535–544.

Katz, B. and Miledi, R. (1967b), *Proc. Roy. Soc. (B),* **167**, 23–38.

Katz, B. and Miledi, R. (1972), *J. Physiol.*, **224**, 665–669.

Katz, B. and Miledi, R. (1973), *J. Physiol.*, **231**, 549–574.

Katz, B. and Thesleff, S. (1957), *J. Physiol.*, **137**, 267–278.

Keesey, J. and Lindstrom, J. (1977), *Excerpta Medica,* **427**, 243–244.

Keesey, J., Lindstrom, J. and Cokely, A. (1977), *New Engl. J. Med.*, **296**, 55.

Kelly, J.J., Lambert, E.H. and Lennon, V.A. (1978), *Ann. Neurol.*, **4**, 67–72.

Kessler, S.W. (1976), *J. Immunol.*, **117**, 1482–1490.

Klymkowsky, M. and Stroud, R. (1979), *J. Mol. Biol.*, **128**, 319–334.

Kuffler, S.W. and Yoshikami, D. (1975), *J. Physiol.*, **251**, 465–482.

Lambert, E.H. and Elmquist, D. (1971), *Ann. N.Y. Acad. Sci.*, **183**, 183–199.

Lambert, E.H., Lindstrom, J.M. and Lennon, V.A. (1976), *Ann. N.Y. Acad. Sci.*, **274**, 300–318.

Lefvert, A.K. and Matell, G. (1979), in *Plasmapheresis and the Immunobiology of Myasthenia Gravis* (Dau, P., ed.), Houghton Mifflin, Boston, pp. 151–160.

Lennon, V.A., Lindstrom, J. and Seybold, M.E. (1975), *J. Exp. Med.*, **141**, 1365–1375.

Lennon, V.A., Lindstrom, J. and Seybold, M.E. (1976), *Ann. N.Y. Acad. Sci.*, **274**, 283–299.

Lennon, V.A., Seybold, M.E., Lindstrom, J., Cochrane, C. and Yulevitch, R. (1977), *J. Exp. Med.*, **147**, 973–983.

Lindstrom, J. (1976), *J. Supramol. Struc.*, **4**, 389–403.

Lindstrom, J. (1977a), *Excerpta Medica, ICS,* **404**, 121–131.

Lindstrom, J. (1977b), *Clin. Immunol. Immunopath,* **7**, 36–43.

Lindstrom, J. (1979), *Advances in Immunology* (Kunkel, H.G. and Dixon, F., eds), Academic Press, New York, Vol. 27.

Lindstrom, J., Anholt, R., Einarson, B., Engel, A., Osame, M. and Montal, M. (1980b), *J. Biol. Chem.*, **255**, 8340–8350.

Lindstrom, J., Campbell, M. and Nave, B. (1978b), *Muscle & Nerve,* **1**, 140–145.

Lindstrom, J., Cooper, J. and Tzartos, S. (1980), *Biochemistry,* **19**, 1454–1458.

Lindstrom, J. and Einarson, B. (1979), *Muscle & Nerve,* **2**, 173–179.

Lindstrom, J., Einarson, B., Lennon, V.A. and Seybold, M.E. (1976b), *J. Exp. Med.*, **144**, 726–738.

Lindstrom, J., Einarson, B. and Merlie, J. (1978a), *Proc. Natl. Acad. Sci. USA,* **75**, 769–773.

Lindstrom, J., Engel, A.G., Seybold, M.E., Lennon, V.A. and Lambert, E.H. (1976c), *J. Exp. Med.*, **144**, 739–753.

Lindstrom, J., Gullick, W., Conti-Tronconi, B. and Ellisman, M. (1980), *Biochemistry,* **19**, 4791–4795.

Lindstrom, J. and Lambert, E.H. (1978), *Neurology*, **28**, 130–138.

Lindstrom, J., Lennon, V., Seybold, M.E. and Whittingham, S. (1976a), *Ann. N.Y. Acad. Sci.*, **274**, 254–274.

Lindstrom, J., Merlie, J. and Yogeeswaran, G. (1979a), *Biochemistry*, **18**, 4465–4470.

Lindstrom, J. and Patrick, J. (1974), *Synaptic Transmission and Neuronal Interaction* (Bennett, M.V.L., ed.), Raven Press, New York, pp. 191–216.

Lindstrom, J., Seybold, M.E., Lennon, V.A., Whittingham, S. and Duane, D.D. (1976a), *Neurology,* **26**, 1054–1059.

Lindstrom, J., Walter, B. and Einarson, B. (1979b), *Biochemistry,* **18**, 4470–4480.

Magleby, K.L. (1979), *Prog. Brain. Res.,* **49**, 175–182.

Marengo, T.S., Harrison, R., Lunt, G.G. and Behan, P.O. (1979), *Lancet,* **1**, 442.

Martin, A.R. (1977), *Handbook of Neurophysiology,* American Physiological Society, Bethesda, MD, Section 1, Vol. 1, pp. 329–355.

McMahan, U.J., Sanes, J.S. and Marshall, L.M. (1978), *Nature,* **271**, 172–174.

Merlie, J., Heinemann, S.J., Einarson, B. and Lindstrom, J. (1979b), *J. Biol. Chem.,* **254**, 6328–6332.

Merlie, J.P., Heinemann, S.J. and Lindstrom, J. (1979a), *J. Biol. Chem.,* **254**, 6320–6327.

Mittag, T., Kornfeld, P., Tormay, A. and Woo, C. (1976), *New Engl. J. Med.,* **294**, 691–694.

Molenaar, P.C., Polak, R.L., Miledi, R., Alema, S., Vincent, A. and Newsom-Davis, J. (1979), *Prog. Brain Res.,* **49**, 449–458.

Monnier, V. and Fulpius, B. (1977), *Clin. Exp. Immunol.,* **29**, 16–22.

Morgutti, M., Conti-Tronconi, B., Sghirlanzoni, A. and Clementi, F. (1979), *Neurology,* **29**, 734–738.

Müller-Eberhard, H.J. (1975), *Annu. Rev. Biochem.,* **44**, 697–724.

Naiem, F., Keesey, J., Herrmann, D., Lindstrom, J., Zeller, E. and Wolford, R. (1978), *Tissue Antigens,* **12**, 381–386.

Namba, T., Brown, S.B. and Grob, D. (1970), *Pediatrics,* **45**, 488–504.

Nastuk, W., Niemi, W., Alexander, J., Chang, H. and Nastuk, M. (1979), *Am. J. Physiol.,* **236**, C53–C57.

Nathanson, N. and Hall, Z. (1979), *Biochemistry,* **18**, 3392–3401.

Newsom-Davis, J., Pinching, A.J., Vincent, A. and Wilson, S. (1978), *Neurology,* **28**, 266–272.

Newsom-Davis, J., Wilson, S., Vincent, A. and Ward, C. (1979), *Lancet,* 464–468.

Oosterhuis, H.G., Feltkamp, T.E. and van Rossum, A.L. (1976), *Ann. N.Y. Acad. Sci.,* **274**, 468–474.

Osserman, K.E. and Genkins, G. (1971), *Mt. Sinai J. Med.,* **38**, 497–537.

Özdemir, C. and Young, R.R. (1976), *Ann. N.Y. Acad. Sci.,* **274**, 203–222.

Padykula, H.A. and Gauthier, G.F. (1970), *J. Cell Biol.,* **46**, 27–41.

Papatestas, A.E., Genkins, G., Horowitz, S.H. and Kornfeld, P. (1976), *Ann. N.Y. Acad. Sci.,* **274**, 555–573.

Patrick, J. and Lindstrom, J. (1973), *Science,* **180**, 871–872.

Patrick, J., Lindstrom, J., Culp, B. and McMillan, J. (1973), *Proc. Natl. Acad. Sci. USA,* **70**, 3334–3338.

Peers, J., McDonald, B. and Dawkins, R. (1977), *Clin. Exp. Immunol.,* **27**, 66–73.

Penn, A., Chang, H., Lovelace, R., Niemi, W. and Miranda, A. (1976), *Ann. N.Y. Acad. Sci.,* **274**, 354–376.

Pinching, A.J., Peters, D.K. and Newsom, J.N. (1976), *Lancet,* **2**, 1373–1376.

Pirskanen, R. (1976), *Ann. N.Y. Acad. Sci.,* **274**, 451–460.

Porter, C.W. and Barnard, E.A. (1975), *Exp. Neurol.,* **48**, 542–556.

Potter, L.T. and Smith, D.S. (1977), *Tissue & Cell,* **9**, 585–644.

Prives, J., Hoffman, L., Tarrab-Hazdai, R., Fuchs, S. and Amsterdam, A. (1979), *Life Sci.*, **24**, 1713–1718.

Raftery, M.A., Vandler, R.L., Reed, K.L. and Lee, T. (1976), *Cold Spring Harbor Symp. Quant. Biol.*, **40**, 193–202.

Reiness, C. and Hall, Z. (1977), *Nature*, **268**, 655–657.

Reiness, G., Hogan, P., Marshall, J., Hall, Z., Griffin, G. and Goldberg, A. (1977), *Cellular Neurobiology* (Hall, Z., Kelley, R. and Fox, C.F., eds), pp. 207–215.

Reuther, P., Fulpius, B.W., Mertens, H.G. and Hertel, G. (1979), in *Plasmapheresis and the Immunobiology of Myasthenia Gravis* (Dau, P., ed.), Houghton Mifflin, Boston, pp. 329–342.

Reynolds, J.A. and Karlin, A. (1978), *Biochemistry*, **17**, 2035–2038.

Richman, D., Antel, J., Patrick, J. and Arnason, B. (1979), *Neurology*, **29**, 291–296.

Richman, D., Patrick, J. and Arnason, B.G.W. (1976), *New Engl. J. Med.*, **294**, 694–698.

Rosenberry, T.R. (1975), *Adv. Enzymol.*, **43**, 103–218.

Ross, M.J., Klymkowsky, M.W., Agard, D.A. and Stroud, R.M. (1977), *J. Mol. Biol.*, **116**, 635–659.

Russell, A. and Lindstrom, J. (1978), *Neurology*, **28**, 847–849.

Sahashi, K., Engel, A.G., Lambert, E.H. and Howard, F.M. (1980), *J. Neuropathol. Exp. Neurol.*, **39**, 160–172.

Sahashi, K., Engel, A.G., Lindstrom, J., Lambert, E.H. and Lennon, V. (1978), *J. Neuropathol. Exp. Neurol.*, **37**, 212–223.

Sakman, B. (1978), *Fed. Proc.*, **37**, 2654–2659.

Salpeter, M., Kasprazak, H., Feng, H., Fertuck, H. (1979), *J. Neurocytology*, **8**, 95–115.

Sanders, D.B., Howard, J.F., Johns, T.R. and Campa, J.F. (1979), in *Plasmapheresis and the Immunobiology of Myasthenia Gravis*, (Dau, P., ed.), Houghton Mifflin, Boston, pp. 289–306.

Santa, T. and Engel, A.G. (1973), in *New Developments in EMG and Clinical Neurophysiology* (Desmedt, J.E., ed.), S. Karger, Basel, Vol. 1, pp. 41–54.

Satyamurti, S., Drachman, D. and Stone, F. (1975), *Science*, **187**, 955–957.

Schrader, P.L., Peters, H.A. and Dahl, D.S. (1972), *Arch. Neurol.*, **27**, 456–457.

Seybold, M.E., Baergen, R., Nave, B. and Lindstrom, J. (1978), *Brit. Med. J.*, **2**, 1051–1053.

Seybold, M.E. and Drachman, D.B. (1974), *New Engl. J. Med.*, **290**, 81–84.

Seybold, M.E., Lambert, E., Lennon, V. and Lindstrom, J. (1976), *Ann. N.Y. Acad. Sci.*, **274**, 275–282.

Seybold, M.E. and Lindstrom, J. (1979), in *Plasmapheresis and the Immunobiology of Myasthenia Gravis* (Dau, P., ed.), Houghton Mifflin, Boston, pp. 307–314.

Seybold, M.E. and Lindstrom, J. (1981), *New York Academy of Sciences*, in press.

Shorr, R.G., Dolly, O. and Barnard, E. (1978), *Nature*, **274**, 283–284.

Simpson, J. (1960), *Scot. Med. J.*, **5**, 419–436.

Sine, S. and Taylor, P. (1979), *J. Biol. Chem.*, **254**, 3315–3325.

Sobel, A., Weber, M. and Changeux, J.P. (1977), *Eur. J. Biochem.*, **80**, 215–224.

Soudag-Tschroots, I., Schultz-Raateland, R., Van Walbeak, H. and Feltkamp, T. (1979), *Clin. Exp. Immunol.*, **37**, 323–327.

Stradler, C., Revel, J.P. and Raftery, M.A. (1979), *J. Cell Biol.*, **83**, 499–510.

Sugiyama, H., Benda, P., Meunier, J. and Changeux, J. (1973), *FEBS Lett.*, **35**, 124–128.

Takeuchi, A and Takeuchi, N. (1960), *J. Physiol.*, **154**, 52–67.

Talal, N. (1977), *Autoimmunity*, Academic Press, 184–207.

Tarrab-Hazdai, R., Aharonov, A., Abramsky, O., Yaar, I. and Fuchs, S. (1975a), *J. Exp. Med.*, **142**, 785–789.

Tarrab-Hazdai, R., Aharonov, A., Silman, I., Fuchs, S. and Abramsky, Ó. (1975b), *Nature*, **256**, 128–130.

Thornell, L.E., Sjöström, M., Mattsson, C.H. and Heilbronn, E. (1976), *J. Neurol., Sci.*, **29**, 389–410.

Tokya, K.V., Birnberger, K.L., Anzil, A.P., Schlegel, C., Besinger, V. and Struppler, A. (1978), *J. Neurol. Neurosurg. Psychiatry*, **41**, 746–753.

Tokya, K.V., Drachman, D.B., Griffin, D.E., Pestronk, A., Winkelstein, J.A., Fischbeck, K.H. and Kao, I. (1977), *New Engl. J. Med.*, **296**, 125–131.

Tokya, K.V., Drachman, D.B., Pestronk, A. and Kao, I. (1975), *Science*, **190**, 397–399.

Tzartos, S. and Lindstrom, J. (1980), *Proc. Natl. Acad. USA*, **77**, 755–759.

Vandlen, R., Wilson, C., Eisenarch, J. and Raftery, M. (1979), *Biochemistry*, **18**, 1845–1854.

Vincent, A. and Newsom-Davis, J. (1979), *Lancet*, 441–442.

Vincent, A., Newsom-Davis, J. and Martin, V. (1978b), *Lancet*, **1**, 1254.

Vincent, A., Scadding, G.K., Clarke, C. and Newsom-Davis, J. (1979), in *Plasmapheresis and the Immubobiology of Myasthenia Gravis* (Dau, P., ed.), Houghton Mifflin, Boston, pp. 59–71.

Vincent, A., Scadding, G.K., Thomas, H. and Newsom-Davis, J. (1978a), *Lancet*, **1**, 305–307.

Weber, M., David-Pfeuty, T. and Changeux, J.P. (1975), *Proc. Natl. Acad. Sci. USA*, **72**, 3443–3447.

Weinberg, C. and Hall, Z. (1979), *Proc. Natl. Acad. Sci. USA*, **76**, 504–508.

Wekerle, H., Patterson, B., Ketelson, V.P. and Feldman, M. (1975), *Nature*, **256**, 493–494.

Wernig, A. (1975), *J. Physiol.*, **244**, 207–221.

Wise, D.S., Karlin, A. and Schoenborn, B. (1979), *Biophys. J.*, **28**, 473–496.

Wolosin, J., Lyddiatt, A., Dolly, J., Barnard, E. (1980), *Eur. J. Biochem.*, **109**, 495–505.

Wu, C.S.W. and Raftery, M. (1979), *Biochem. Biophys. Res. Commun.*, **89**, 26–35.

Zucker, R.S. (1973), *J. Physiol.*, **229**, 787–810.

9   Thyrotropin Receptor Antibodies

BERNARD REES SMITH

Receptor Regulation
(*Receptors and Recognition*, Series B, Volume 13)
Edited by R. J. Lefkowitz
Published in 1981 by Chapman and Hall, 11 New Fetter Lane, London EC4P 4EE
© 1981 Chapman and Hall

9.1 INTRODUCTION

The sera of patients with hyperthyroid Graves' disease contain antibodies to the thyrotropin (TSH) receptor. These antibodies appear to induce hyperthyroidism in Graves' disease by binding to the receptor and mimicking the action of TSH. This review considers the properties of the antibodies and their role in the disease process.

9.2 DISCOVERY OF TSH-RECEPTOR ANTIBODIES

In 1956, Adams and Purves observed that the serum of patients with Graves' disease contained thyroid-stimulating activity quite distinct from thyrotropin (TSH). Because of the prolonged time course of action of the abnormal stimulator in the bioassay systems used to measure it the stimulator was referred to as Long-Acting Thyroid-Stimulator (LATS) (Dorrington and Munro, 1966). LATS was subsequently shown to be an immunoglobulin of the IgG class with all the characteristics of a thyroid-stimulating autoantibody (Dorrington and Munro, 1966; Smith *et al.*, 1969a). Later other thyroid-stimulating autoantibodies with slightly different properties from LATS were discovered (Adams and Kennedy, 1967, 1971; Dirmikis and Munro, 1975). The effects of LATS on the thyroid appeared to be virtually identical to those of TSH with the actions of both types of substances (Dorrington and Munro, 1966) being mediated by the adenylate cyclase–cyclic AMP system (Kendall-Taylor, 1972, 1973). It was clearly possible therefore that TSH and the autoantibodies in Graves' sera stimulated thyroid function by interacting with the same cell surface receptor. Evidence for this concept was first provided by studies of the effects of Graves' IgG on ^{125}I-labeled TSH binding to its receptor in thyroid membranes. These investigations indicated that TSH binding was readily inhibited in a dose-dependent manner by Graves'-IgG and suggested that Graves' IgG contained antibodies to the TSH receptor (Manley *et al.*, 1974; Smith and Hall, 1974; Mehdi and Nussey, 1975).

9.3 GENERAL PROPERTIES OF TSH-RECEPTOR ANTIBODIES

9.3.1 Mechanism of the effect of Graves' IgG on TSH binding to thyroid membranes

The ability of Graves' IgG to inhibit TSH binding to thyroid membranes could result from interaction between the IgG and TSH or an effect of the Graves' IgG on thyroid membranes.

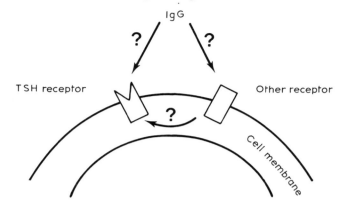

Solve problem by studying solubilized isolated receptors

Fig. 9.1 Possible mechanisms for the effects of Graves' IgG on the TSH binding to thyroid membranes.

Studies using gel filtration and polyethylene glycol precipitation have failed to show any interaction between TSH and Graves' IgG. Furthermore, when thyroid membranes are incubated with Graves' IgG and then washed prior to addition of labeled TSH, inhibition of hormone binding is still observed (Davies, 1977). Consequently, the Graves' IgG must act by binding to the membrane and not by interacting with TSH.

The Graves' IgG could possibly interact directly with the TSH receptor. Alternatively, binding to a different membrane component could occur and this process could somehow result in transmission of a signal through the membrane which inactivated the TSH binding site (Fig. 9.1). One way of deciding between these two possibilities is to disperse the membrane components in an excess of non-ionic detergent micelles and re-examine the effects of Graves' IgG on TSH binding. When this is carried out, the Graves' IgG is found to inhibit TSH binding to the detergent solubilized receptors in a very similar manner to TSH binding to membrane-bound receptors (Manley *et al.*, 1974; Petersen *et al.*, 1977). Consequently, the Graves' IgG appears to interact directly with a component of the TSH receptor and it would seem, therefore, that it contains antibodies to the TSH receptor.

9.3.2 TSH-receptor antibodies and thyroid stimulation

As Graves' IgG contains antibodies to the TSH receptor, a likely mechanism for the thyroid-stimulating action of Graves IgG is that these antibodies act as TSH agonists. This hypothesis is very difficult to prove absolutely at the

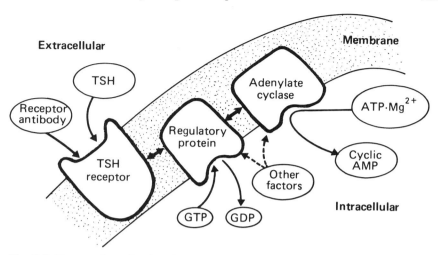

Fig. 9.2 Proposed mechanism by which TSH and TSH-receptor antibodies stimulate the thyroid cell. (Modified from Baxter and Funder, 1979.)

present time as it is always possible to propose that the IgG preparations also contain another antibody which stimulates the thyroid cell by interacting with another receptor. If a non-TSH receptor is involved, however, this has to be a rather specialized molecule able to undergo a complex coupling reaction with adenylate cyclase (Baxter and Funder, 1979). The simplest explanation for the thyroid-stimulating properties of Graves' IgG is to attribute them to TSH-receptor antibodies acting as TSH agonists (Fig. 9.2). This simple hypothesis provides a ready explanation for all the known properties of Graves' IgG and will be elaborated in more detail subsequently.

9.3.3 TSH-receptor antibody levels in different groups of patients

Initial studies suggested that nearly 100 per cent of patients with Graves' disease contained detectable levels of TSH-receptor antibodies (Mukhtar *et al.*, 1975). However, these observations were made on patients selected on the basis of age (over 45 years) because of their inclusion in ^{131}I uptake studies, and subsequent studies on 150 unselected Graves' patients in the Newcastle area indicate that about 75 per cent have detectable serum TSH-receptor antibodies (Fig. 9.3). A recent report from Hong Kong indicates that 84 per cent of Chinese patients with Graves' disease (Teng and Yeung, 1980) have detectable receptor antibodies. Lower incidences have been reported in other areas (Docteur *et al.*, 1980; O'Donnell *et al.*, 1978; Kuzuya *et al.*, 1979; Schleusener *et al.*, 1978).

These variations must reflect several factors including geographical location, clinical and biochemical criteria used to define Graves' disease, patient selection and assay methodology.

About 10 per cent of patients with Hashimoto's thyroiditis have detectable TSH-receptor antibodies (Fig. 9.3) (Mukhtar *et al.*, 1975) and in some cases these can act as TSH agonists and stimulate thyroid function (Clague *et al.*, 1976) (hyperthyroidism does not occur because of extensive autoimmune destruction). The absence of detectable receptor antibody levels in 90 per cent of Hashimoto patients provides good evidence for the specificity of the receptor assay system. High titers of antibodies to thyroglobulin and thyroid microsomes do not influence TSH binding to its receptor (Smith and Hall, 1974).

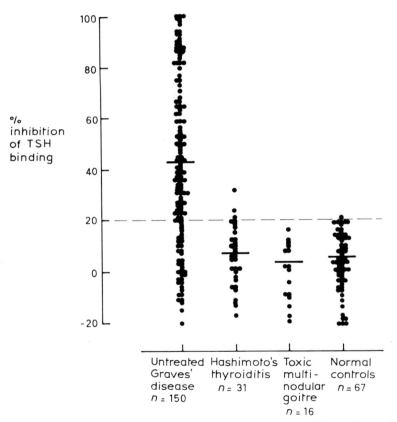

Fig. 9.3 Inhibition of labeled TSH binding to human thyroid membranes by immunoglobulins from the sera of different groups of patients. Values of greater than 20 per cent inhibition, can be considered to represent detectable TSH-receptor antibody activity.

There has been one report of TSH-receptor antibody activity in patients with toxic multinodular goitre (Brown *et al.*, 1978). In our own experience (Fig. 9.3) (McGregor *et al.*, 1979a) and those of other laboratories (Bolk *et al.*, 1979) this does not appear to be the case and the reasons for the discrepancy are not clear.

Assays for TSH-receptor antibody activity are usually carried out with concentrates of serum immunoglobulins. Unfortunately, normal serum immunoglobulins non-specifically inhibit labeled TSH binding to a certain extent (Hall *et al.*, 1975) and the effects of test immunoglobulins must always be considered relative to those of immunoglobulins from an appropriate normal control population.

It is possible that serum TSH could interfere in the measurements of receptor antibody activity. However, serum levels of 100 μU per ml or less do not influence the effect of immunoglobulins on TSH binding to thyroid membranes, but it is advisable to consider serum TSH activity when assessing the results of receptor assays. Thyroglobulin has also been reported to inhibit TSH binding to thyroid membranes (Hashizume *et al.*, 1978) but in our own experience, the concentrations found in serum even after destructive therapy to the thyroid or in thyroid cancer do not influence the effects of serum immunoglobulins on TSH binding.

9.3.4 TSH-receptor antibodies and the control of thyroid function in Graves' disease

In normal subjects, thyroid function is under the control of a feedback system involving pituitary TSH and thyroid hormones (Hall *et al.*, 1975). Perturbation of the system by administration of thyroid hormones or thyrotropin-releasing hormone can be used to assess the state of the pituitary-thyroid axis and the effects of TSH-receptor antibodies (Hall *et al.*, 1975). When this is carried out in patients who have been treated, the presence of receptor antibodies correlates well with the loss of TSH control of thyroid function (Fig. 9.4) (Clague *et al.*, 1976; Kuzuya *et al.*, 1979) suggesting that when the antibodies are present they are involved in thyroid stimulation.

The relationship between levels of receptor antibodies and the extent of thyroid stimulation is difficult to assess as Graves' thyroid tissue is usually subject to some autoimmune destruction as well as autoimmune stimulation. However, the iodine-trapping mechanism is not markedly influenced by autoimmune destruction and consequently early ^{131}I-uptake measurements are probably one of the most useful assessments of the ability of receptor antibodies to stimulate the thyroid. Indeed, in untreated Graves' patients (carefully selected for absence of iodine contamination in their diets) levels of receptor antibodies are found to show a significant correlation with the ^{131}I uptake (Fig. 9.5) (Mukhtar *et al.*, 1975).

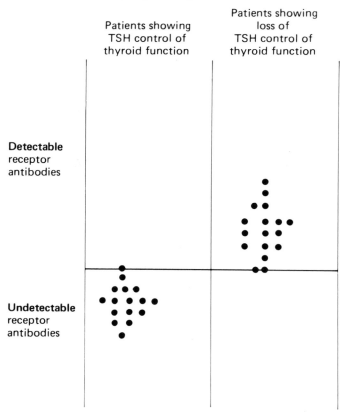

Fig. 9.4 Relationship between the presence or absence of TSH-receptor antibody activity and the control of thyroid function in 33 patients treated for Graves' disease (see Teng *et al.*, 1977). TSH control of thyroid function was assessed by using both T_3 suppression and TRH-stimulation tests (Hall *et al.*, 1975).

In addition, Endo *et al.* (1978) have shown that levels of TSH-receptor antibodies correlate significantly with 30 min ^{99m}Tc uptake by the thyroid in untreated Graves' patients. These investigators also demonstrated a good relationship between the levels of receptor antibodies and the extent of thyroid follicle hyperplasia in needle biopsy samples.

These initial observations on the relationship between receptor antibody levels and control of thyroid function suggested that when the antibodies were present, they stimulated the thyroid. Consideration of the complexity of the immune response and the heterogeneity of antibody affinity and specificity, however, might suggest that some receptor antibodies could act as TSH

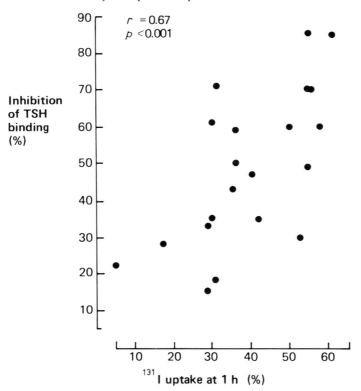

Fig. 9.5 Relationship between TSH-receptor antibody activities and thyroidal
[131]I uptake (one hour after administration of ^{131}I) in 21 patients with untreated
Graves' disease.

antagonists as well as TSH agonists. Subsequently, good evidence for the
existence of receptor antibodies which were unable to stimulate thyroid function was obtained in a few patients with Hashimoto's thyroiditis (Clague *et al.*,
1976) and Ophthalmic Graves' disease (Fig. 9.6) (Teng *et al.*, 1977) (eye
signs of Graves' disease in the absence of hyperthyroidism or a past history of
hyperthyroidism). More recently further support for these observations has
been provided by a report of a euthyroid mother with high levels of receptor antibody (which crossed the placenta) giving birth to a euthyroid child
(Hales *et al.*, 1980). It is also possible that under some circumstances, receptor antibodies acting as TSH antagonists could induce hypothyroidism and
again there has recently been some evidence that this can occur in the neonate
(Konishi *et al.*, 1980).

As mentioned earlier in this section, Graves' disease usually has the fea-

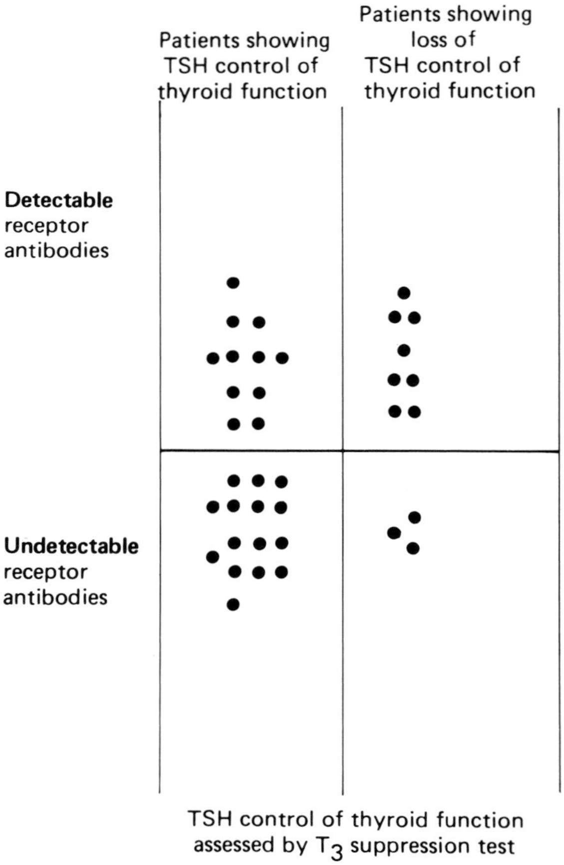

Fig. 9.6 Relationship between the presence or absence of TSH-receptor antibodies and control of thyroid function in 37 patients with ophthalmic Graves' disease.

tures of both autoimmune stimulation and autoimmune destruction. In hyperthyroid Graves' disease the stimulatory process is clearly predominant. In some patients with ophthalmic Graves' disease, however, there appears to be a stable balance between autoimmune stimulation and destruction (Teng *et al.*, 1977).

A further factor in the process of thyroid stimulation by TSH-receptor antibodies is the ability of thyroid tissue to become refractory to stimulation (Kendall-Taylor, 1973; Shuman *et al.*, 1976).

Overall, therefore, thyroid stimulation by receptor antibodies is a complex process summarized in Fig. 9.7.

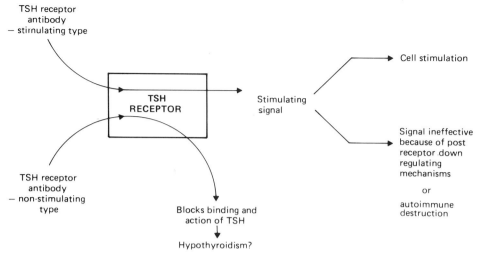

Fig. 9.7 Summary of the effects of two different types of TSH-receptor antibodies.

Receptor antibodies can be of both TSH agonist and TSH antagonist types and most Graves' sera probably contain a mixture of antibodies with agonist, partial agonist and antagonist properties. In our own experience, the agonist or stimulating properties virtually always predominate and induce a stimulating signal in the receptor. This stimulating signal usually leads to thyroid stimulation and hyperthyroidism. Sometimes, however, the stimulating signal is rendered ineffective or at least attenuated by autoimmune destruction and tissue refractoriness.

9.3.5 Measurement of the thyroid-stimulating activity of Graves' IgG

The interaction between Graves' IgG and the thyroid is frequently assessed in terms of a stimulatory response of various thyroid preparations. A particularly useful method appears to be that of Onaya *et al.* (1973) which depends on the production of cyclic AMP in human thyroid slices. Several laboratories have shown that this technique is slightly more sensitive than the radioreceptor assay technique in so far as the sera from a few patients who appear to have undetectable receptor antibody activity are able to stimulate cyclic AMP production (Kuzuya *et al.*, 1979; Sugenoya *et al.*, 1979).

9.3.6 Interaction of Graves' IgG with non-thyroidal tissues

Graves' IgG has been shown to stimulate lipolysis in isolated fat-cells (Kendall-Taylor and Munro, 1971; Hart and McKenzie, 1971). In addition, fat-cells contain readily demonstrable TSH receptors (Teng *et al.*, 1975; Gill *et al.*, 1977) and are responsive to TSH. These observations provide strong circumstantial support for the suggestion that the stimulatory properties of Graves' IgG are due to the effects of TSH-receptor antibodies. If thyroid stimulation is due to the effects of an antibody interaction with a membrane component other than the TSH receptor this hypothesis would have to propose that this membrane component is also present on fat-cells.

Thyrotropin receptors (quite distinct from gonadotropin receptors) have also been demonstrated in the testis (Davies *et al.*, 1978; Amir *et al.*, 1978) and TSH binding to testicular tissue is readily inhibited by TSH-receptor antibodies (Davies *et al.*, 1978). There is as yet no evidence for a physiological or pathophysiological role for the interaction of TSH and TSH-receptor antibodies with the testis.

9.4 SITE OF SYNTHESIS OF TSH-RECEPTOR ANTIBODIES

In this section, Hashimoto's thyroiditis as well as Graves' disease will be considered. Hashimoto's disease is characterized by the presence of serum autoantibodies to thyroglobulin and thyroid microsomes and extensive autoimmune destruction.

Lymphocytic infiltration of the thyroid is a feature of both autoimmune disorders and the immune cells in Hashimoto thyroid tissue show many of the organizational features of lymph nodes (Söderström and Björklund, 1974). In Graves' disease, lymphocytic infiltration is less extensive but aggregates of lymphoid cells are a frequent finding (Swanson Beck *et al.*, 1973).

It is quite possible that the lymphocytes localized in the thyroid are involved in the synthesis of thyroid autoantibodies in Graves' disease and Hashimoto's thyroiditis (Fig. 9.8). In the case of Hashimoto's disease there is some direct evidence for this possibility (Davoli and Salabe, 1976) including studies which have demonstrated that lymphocytes extracted from Hashimoto thyroid tissue synthesize relatively large amounts of thyroglobulin and microsomal antibodies (McLachlan *et al.*, 1979). With regard to Graves' disease, there is considerable indirect evidence that thyroid lymphocytes are also a major site of TSH-receptor antibody production. This evidence comes principally from studies of the effects of treatment for Graves' disease on serum levels of TSH-receptor antibodies and other thyroid antibodies (about 80 per cent of patients with Graves' disease have serum microsomal antibodies detectable by the tanned red cell haemagglutination inhibition technique).

If lymphocytes localized in the thyroid are a major site of autoantibody synthesis, antithyroid treatment for Graves' hyperthyroidism might be expected to influence autoantibody production by thyroid lymphocytes as well as thyroid hormone synthesis by thyroid tissue.

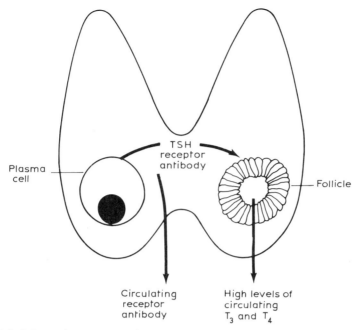

Fig. 9.8 Schematic representation of TSH-receptor antibody synthesis by lymphocytes localized in the thyroid of a patient with Graves' disease.

Three types of antithyroid treatment are generally used in Graves' hyperthyroidism: (1) antithyroid drugs such as carbimazole or propylthiouracil which are concentrated by the thyroid and block iodine organification; (2) destructive therapy with ^{131}I; (3) removal of a large part of the thyroid by surgery (subtotal thyroidectomy). In addition, treatment directed against the effects of thyroid hormones at the tissue level is also used in the form of the β-blocking drug propranolol.

These different forms of therapy are followed in most, but not all, patients by characteristic changes in serum levels of microsomal and TSH-receptor antibodies.

(1) Antithyroid drug therapy is accompanied by a marked fall in both autoantibodies (Mukhtar *et al*., 1975; Teng and Yeung, 1980; McGregor *et al*., 1978, 1979a, 1980a,b; Pinchera *et al*., 1969; Einhorn *et al*., 1967; Fenzi *et al*., 1979), and this appears to be independent of changes in thyroid status such as serum thyroid hormone levels.

(2) [131]I treatment is followed by a marked increase in thyroid autoantibody synthesis (Fenzi *et al*., 1979; McGregor *et al*., 1978, 1979a). Serum levels are maximal about three months after therapy and then fall away over the succeeding months.

(3) The effects of surgery are more difficult to analyse because most patients are treated with antithyroid drugs until immediately before thyroidectomy. However, in patients who have detectable microsomal or TSH-receptor antibodies, serum levels tend to fall after surgery and remain low or undetectable (Mukhtar *et al.*, 1975; McGregor *et al.*, 1980a).

(4) Treatment of patients for short periods (two months) with propranolol or placebo is not accompanied by any significant change in serum autoantibody levels (McGregor *et al.*, 1980b).

These observations indicate that: (a) spontaneous changes in autoantibody levels do not tend to occur frequently over short periods of time; (b) antithyroid drugs inhibit autoantibody synthesis directly or indirectly; (c) irradiation of the thyroid transiently stimulates autoantibody production; (d) removal of a large proportion of the thyroid results in reduced thyroid autoantibody synthesis.

Studies *in vitro* have demonstrated that concentrations of antithyroid drugs similar to those found in the thyroid during treatment readily inhibit autoantibody synthesis by lymphocyte cultures (McGregor *et al.*, 1980a,b). Furthermore, lymphocytic infiltration of the thyroid is markedly reduced during antithyroid drug treatment (Swanson Beck *et al.*, 1973). Consequently antithyroid drugs concentrated in the thyroid and acting directly on thyroid lymphocytes to inhibit autoantibody production appears to be a likely explanation for the fall in serum autoantibody levels during treatment.

Similarly, irradiation of lymphocytes *in vitro* can be shown to generate a population of cells capable of stimulating autoantibody synthesis (McGregor *et al.*, 1979b) and such a mechanism could be important in the increase in autoantibody synthesis observed after irradiation of the thyroid. With regard to surgery, the removal of a large proportion of the lymphocytes synthesizing thyroid autoantibodies would be expected to result in a marked reduction in serum levels.

The role of antigen release which probably occurs during ^{131}I treatment and surgery on autoantibody synthesis is not clear. Both procedures probably result in similar marked elevations in concentrations of autoantigen as evidenced by estimations of serum thyroglobulin levels (Van Herle *et al.*, 1979). However, the changes in serum autoantibody levels are opposite following the two procedures (increase after ^{131}I treatment, decrease after surgery) and this suggests that antigen release does not have a major role in the acute changes in autoantibody levels observed following surgery or ^{131}I treatment. In addition, stimulation of autoantibody production *in vitro* by irradiated lymphocytes can be demonstrated to occur in the absence of autoantigen (McGregor *et al.*, 1979b).

In summary, therefore, the evidence for lymphocytes localized in the thyroid being a major site of autoantibody production is as follows.

(1) Relatively large amounts of microsomal antibodies are produced by lymphocytes isolated from Hashimoto thyroid tissue.

(2) Microsomal- and TSH-receptor antibody synthesis in Graves' disease are influenced by antithyroid treatment in a parallel fashion.

(3) The effects of antithyroid drugs, ^{131}I and surgery on microsomal- and TSH-receptor antibody production are consistent with the autoantibodies being synthesized by thyroid lymphocytes.

9.5 THE TSH RECEPTOR

9.5.1 Properties

Analysis of the TSH binding characteristics of non-ionic detergent-solubilized thyroid membranes with enzymatic, chemical and affinity probes indicates that the porcine TSH receptor is an amphiphilic membrane glycoprotein with at least one essential inter- or intra-chain disulfide bridge (Brennan *et al.*, 1980; Ginsberg *et al.*, 1980). When solubilized in detergents such as Triton or Lubrol the complex formed between TSH, the TSH-receptor and detergent micelles has a molecular weight a little less than 300 000 (Dawes *et al.*, 1978). A rough correction for the contribution made by hormone and micelle suggests that the receptor itself has a molecular weight in the region of 150 000. The isoelectric point of the detergent-solubilized receptor is about 4.5 (Dawes *et al.*, 1978).

The association constant of the TSH–TSH-receptor interaction is in the region of 10^9 molar^{-1} (Manley *et al.*, 1974; Smith and Hall, 1974; Mehdi and Nussey, 1975; Dawes *et al.*, 1978).

Detergent-solubilized TSH-receptor preparations are only stable for a few days at 4° C and for a few hours at 20° C (Dawes *et al.*, 1978). In addition, the receptors are present in very small amounts (Manley *et al.*, 1974; Smith and Hall, 1974; Mehdi and Nussey, 1975; Dawes *et al.*, 1978) and several hundred kilograms of thyroid tissue appear to be required to prepare mg amounts of pure receptor. Consequently, even if solubilized TSH-receptor preparations can be stabilized (for example by introducing phospholipids into the detergent micelles) (Agnew and Raftery, 1979) complete purification is a considerable challenge with currently available technology. Developments in genetic engineering which could be used to increase greatly the rate of TSH-receptor synthesis may be required before satisfactory receptor purification can be achieved.

9.5.2 The interaction of TSH-receptor antibodies with the receptor

The regions of the TSH receptor which are involved in receptor antibody binding are unknown at present. A molecule as large as the receptor would be expected to have multiple antigenic sites and Graves' IgG probably contains antibodies to several different antigenic determinants on the receptor. In this respect, however, it is of interest that the Fab fragments of receptor antibodies are equally effective in inhibiting TSH binding as the intact IgG (Hall *et al.*, 1975; Rees Smith *et al.*, 1977). Furthermore, the binding of TSH and TSH-receptor antibodies to the TSH receptor appear to be mutually exclusive in so far as there is no evidence for the formation of a complex consisting of all three components (Manley *et al.*, 1974; Petersen *et al.*, 1977; Rees Smith and Hall, 1980). It would appear, therefore, that although the receptor antibodies may not bind directly to the TSH binding site on the receptor the antibody binding sites are closely linked to the hormone binding site.

9.5.3 The TSH receptor and LATS-absorbing activity

Freezing and thawing of human thyroid particulate preparations has been shown to release a water-soluble substance which will reversibly bind and inactivate the LATS activity of Graves' IgG (Smith, 1970; Dirmikis and Munro, 1973). This material is sometimes referred to as LATS absorbing activity or LAA. LAA has a molecular weight in the region of 30 000 and isoelectric point of about 4.5 (Smith, 1971). It has not so far been possible to demonstrate direct binding of TSH to LAA (Dawes *et al.*, 1978) although this may reflect an ability to separate LAA bound and free TSH.

The relationship between LAA and the TSH receptor is not clear at present. It would seem likely, however, that LAA is a hydrophilic fragment of the amphiphilic TSH receptor cleaved off by enzymes released during freezing and thawing of thyroid tissue (Dawes *et al.*, 1978).

9.5.4 LATS protector

The binding of LATS to LAA can be inhibited by another antibody and this antibody, which appears to compete with LATS for LAA, is described as LATS protector (LATS-P) (Nutt *et al.*, 1974; Adams *et al.*, 1976; Dirmikis *et al.*, 1974; Solomon *et al.*, 1977). The properties of LATS-P in Graves' sera have been studied in some detail. In brief, LATS-P activity is present in the serum of most patients with hyperthyroid Graves' disease (Adams *et al.*, 1976; Solomon *et al.*, 1977) and the level of activity correlates with thyroid-stimulation as assessed by iodine clearance rates (Adams *et al.*, 1976). Furthermore, the presence of high levels of LATS-P in the serum of pregnant women has been shown to be a valuable predictor of neonatal hyperthyroidism (Munro *et al.*, 1978). However, as with antibody activity measured by inhibition of TSH binding, biologically inactive LATS-P has been observed in the serum of some patients (Ozawa *et al.*, 1979).

9.5.5 The relationship between *in vitro* measurements of thyroid-stimulating and TSH-receptor binding properties of Graves' IgG

Preliminary studies in a small number of patients suggested that the ability of Graves' IgG to inhibit TSH binding to human thyroid membranes correlated with their ability to stimulate cyclic AMP production in the same membrane preparations (Mukhtar *et al.*, 1975) but insufficient patients were studied to make a statistical analysis.

Subsequently more extensive observations were made in which TSH-receptor antibody levels in Graves' IgG were compared with the ability of the IgG to stimulate cyclic AMP production by thyroid slices (Kuzuya *et al.*, 1979; Sugenoya *et al.*, 1979). Overall, no significant correlation was observed between the two measurements but the validity of the statistical analysis can be called into question. In particular the assignment of specific values to measurements made in the undetectable range of each assay would seem inappropriate. If the data is analysed from a slightly different aspect and only samples having detectable receptor antibodies and detectable slice-stimulating activity are compared a correlation would appear to exist.

Overall, comparison of the thyroid slice-stimulating properties of Graves' IgG with the TSH-receptor binding properties would suggest that when both activities are detectable, the levels of the two activities show a rough correlation. In addition, binding activity is sometimes found in the absence of stimulation. A few Graves' IgGs show some stimulation in the absence of detectable receptor binding and this probably reflects the greater sensitivity of the biological response. These observations are quite consistent with the hypothesis that Graves' IgG contain a heterogenous population of receptor antibodies with a mixture of TSH agonist, partial agonist and antagonist properties.

9.5.6 TSH-receptor antibodies and LATS activity

LATS activity is measured conventionally in a bioassay system which depends on ^{131}I release from the mouse thyroid *in vivo* (Dorrington and Munro, 1966; McKenzie, 1958a,b). As readily detectable levels of LATS activity are present in a relatively small proportion of Graves' sera detailed analysis of the relationship between levels of LATS activity and other properties of Graves' IgG is difficult. However, when LATS activity is detectable, TSH-receptor antibody activity is also detectable (Clague *et al.*, 1976) and the properties of LATS are consistent with this antibody acting by way of the TSH receptor. Furthermore, a recent report has shown a highly significant correlation between LATS levels and inhibition of TSH binding in 23 patients who were both LATS and receptor antibody positive (Ozawa *et al.*, 1979).

9.5.7 TSH-receptor antibodies and LATS-protector

As LATS-protector competes with LATS for a water-soluble binding component (LATS absorbing activity or LAA) it seems likely that LATS-P binds to LAA, but not necessarily at the same site as LATS. It seems likely that LAA is a component of the TSH receptor (see Section 9.5.3) and consequently

LATS-P activity would appear to be due to an antibody to a component of the TSH receptor.

The ability of Graves' IgG to inhibit TSH binding to thyroid membranes and act as LATS-protector has been compared. Although two of these studies involving 26 patients (Rees Smith *et al.*, 1975) and 29 patients (Ozawa *et al.*, 1979) failed to find a significant correlation between levels of LATS-P and TSH binding inhibition a third and more detailed study (Davies, 1977) of 47 patients showed a significant correlation ($r = 0.43$; $P < 0.01$).

Although LATS, LATS-P and TSH appear to interact with the same receptor molecule they may possibly bind to different sites on the receptor with different affinities. Consequently the relationship between inhibition of TSH binding, LATS activity and LATS-P might be expected to be complex.

9.6 PHYLOGENETIC SPECIFICITY OF AUTOANTIBODY ACTIVITIES IN GRAVES' IgG

From considerations of the heterogeneity of the autoimmune response and the nature of hormone receptors the interaction of TSH-receptor antibodies with thyroid tissue from different species might be expected to show some interesting differences.

The first examples of phylogenetic specificity of this type of antibody was provided by Adams and Kennedy in their discovery of LATS-protector (see Section 9.5.4). These antibodies were able to block LATS activity (measured by mouse bioassay) binding to human thyroid tissue, but were unable to stimulate the mouse thyroid themselves. Subsequently, Wong and Doe (1972) demonstrated potent LATS activity measured by mouse bioassay in the serum of a patient who clearly showed TSH control of thyroid function indicating a discrepancy between ability of the antibodies to stimulate the thyroids of mouse and man. More extensive studies have suggested that LATS and LATS-P activities represent different populations of receptor antibodies with LATS tending to show a preferential ability to stimulate the mouse thyroid and LATS-P a preferential tendency to stimulate the human thyroid (Adams *et al.*, 1976).

A detailed analysis has also been made of the ability of Graves' IgG to stimulate cyclic AMP production by slices of thyroid tissue obtained from different species (McKenzie and Zakarija, 1976; Zakarija and McKenzie, 1978a,b,c). Although all the IgG preparations stimulated human thyroid tissue only a proportion of the preparations were active in thyroid tissue from other species.

This type of specificity could possibly have reflected the different abilities of some antibodies to bind to TSH receptors from different species. However, although preliminary experiments indicated some variations in the ability of different Graves' IgGs to inhibit TSH binding to TSH receptors from different species (Hall *et al*., 1975) these relatively small variations were insufficient to explain the marked species differences observed in stimulatory responsiveness. It would appear therefore that the inability of Graves' IgG with human thyroid-stimulating properties to stimulate thyroid tissue from other species principally represents an inability of the bound antibody to induce receptor coupling with adenylate cyclase.

9.7 SUMMARY OF THE RELATIONSHIP BETWEEN DIFFERENT AUTOANTIBODY ACTIVITIES IN GRAVES' IgG AND CONSIDERATION OF TERMINOLOGY

Analysis of Graves' IgG by different techniques indicates that it can contain a variety of antibodies which (1) stimulate human thyroid tissue, (2) stimulate mouse thyroid tissue (LATS), (3) stimulate thyroids from other non-human species, (4) block the binding of LATS to human thyroid tissue (LATS-P) and (5) inhibit TSH binding to the TSH receptor. In addition, microsomal and thyroglobulin autoantibodies are often present.

With the exception of microsomal and thyroglobulin antibodies which do not cross react with any of the activities summarized above, the other antibody activities are clearly closely related. Current observations are consistent with the hypothesis that all the different antibody activities are directed towards components of the TSH receptor and all could be considered to be TSH-receptor antibodies conveniently abbreviated to TRAb. However, it is important to recognize that the different properties of these antibodies and antibody activities, assessed in stimulating assays, could be referred to as thyroid-stimulating antibodies or TSAb with an appropriate prefix to indicate species, e.g. hTSAB for stimulating human tissue.

LATS-protector clearly presents a terminological problem with no obvious alternative for the type of activity measured. An appropriate modification would be LATS-PAb to emphasize the antibody nature of the material.

9.8 CLINICAL VALUES OF TSH-RECEPTOR ANTIBODY MEASUREMENTS

Diagnosis of hyperthyroid Graves' disease is usually carried out on the basis of clinical signs and various biochemical assessments including serum thyroid hormone levels (Hall *et al*., 1975). Measurement of TSH-receptor antibody activity either by receptor binding assay or thyroid-stimulating assay would appear to be unnecessary in order to make the initial diagnosis. However, receptor antibody determinations can be of value in the diagnosis of ophthalmic Graves' disease (Hall and Rees Smith, 1979). Furthermore, measurements of these antibodies, particularly by stimulating assay, would be expected to be of considerable value in predicting the occurrence of neonatal Graves' disease (see Section 9.5.4).

Prediction of disease course following an initial treatment schedule with antithyroid drugs also appears to be possible in some patients from measurements of receptor antibody levels by TSH-receptor assay immediately prior to drug withdrawal as patients with high levels of receptor antibodies tend to relapse (Davies *et al*., 1977; Teng and Yeung, 1980).

McGregor *et al*. (1980c) have recently improved their success in predicting disease course following carbimazole treatment to virtually 100 per cent by using a combination of receptor antibody determinations and HLA assessment. Patients who were HLA DR3 positive relapsed whereas patients who were DR3 negative but showed detectable levels of receptor antibodies also relapsed.

9.9 BIOCHEMICAL CHARACTERISTICS OF TSH-RECEPTOR ANTIBODIES

Biochemical analysis of TSH-receptor antibodies has been carried out using both receptor binding and thyroid-stimulating systems to assess antibody activity.

Studies using the mouse bioassay demonstrated that the thyroid-stimulating activity of Graves' serum was exclusively associated with the IgG fraction (Benhamou-Glynn *et al*., 1969; Maisey, 1972). Furthermore, the activity was formed by combination of heavy and light chains in the Fab part of the IgG molecule (Dorrington and Munro, 1966; Smith *et al*., 1969a). Isolated heavy chains also showed a small amount of thyroid-stimulating activity (Smith *et al*., 1969a). Ion-exchange (Smith *et al*., 1969b) and isoelectric-focusing (Smith, 1971; Adlkofer *et al*., 1973; Lonergan *et al*., 1973) studies demonstrated that the thyroid-stimulating activity of Graves' IgG showed a wide range of isoelectric points indicating that the antibodies were of polyclonal origin. Studies with immunoglobulin light-chain antisera have also demonstrated that the stimulating activity is distributed approximately equally between antibody molecules with kappa and lambda light chains (Kriss, 1968; Ochi and De Groot, 1968; Maisey, 1972). In addition, stimulating activity in one Graves' serum has been shown to be associated with IgG subclasses 1, 2 and 4, but not IgG 3 (Ochi *et al*., 1977).

Analysis of the properties of Graves' IgG using assay systems other than the mouse bioassay have been less extensive. Stimulation of human thyroid adenylate cyclase has been demonstrated with the Fab fragment of Graves' IgG (Rees Smith *et al*., 1977b) and the activity has been shown to be polyclonal (Zakariya and McKenzie, 1978a). Similarly, the ability of Graves' IgG to inhibit TSH binding to thyroid membranes has been localized in the Fab fragment (Hall *et al.*, 1975; Rees Smith *et al.*, 1977b).

The property of antibodies to the receptor to stimulate the thyroid rather than induce cell lysis by way of the complement system is of interest. However, as the number of receptors on the surface of the thyroid cell is small (in the region of 10^3 sites per cell) and the receptor antibodies are of the IgG class, complement activation might not be expected to occur.

The binding of Graves' IgG to thyroid preparations is not readily reversible (Smith and Munro, 1970) in contrast to the easily dissociable characteristics of the TSH–TSH-receptor complex (Manley *et al.*, 1974; Verrier *et al.*, 1974; Rees Smith *et al.*, 1977a). However, dissociation of antibody–receptor complexes can be affected by treatment with the chaotropic agent sodium thiocyanate (Smith, 1971; Smith and Munro, 1970) and this suggests that hydrophobic bonding is a major factor in the antibody–receptor interaction.

9.10 THE CYTOCHEMICAL ASSAY FOR THYROID STIMULATORS

Bitensky *et al.* (1974) have developed a highly sensitive assay for thyroid stimulators which is based on the ability of these substances to increase the permeability of thyroid follicular cell lysomes. The increase in permeability is measured by using a chromogenic substrate for the enzyme naphthylamidase and the resulting colour changes are monitored by scanning and integrating microdensitometry. The method is extremely demanding technically and only a few laboratories have been able to use the method successfully. However, some valuable results have been obtained with the assay (Petersen *et al.*, 1975a,b) which has a sensitivity for TSH of some 100 000 times that of the radioimmunoassay. Serum TSH levels can be readily measured directly in thyrotoxic patients. In addition, TSH and thyroid-stimulating antibodies can be localized in the 4S and 7S fractions respectively of the same serum sample. Immunoglobulins from normal subjects are not found to be active in the assay and this provides good evidence that antibodies interacting with the TSH receptor are not present in normal individuals.

9.11 PROCESSES INVOLVED IN THE INITIATION OF TSH-RECEPTOR ANTIBODY SYNTHESIS

9.11.1 General features

The problem consists of explaining the appearance of a heterogenous population of IgG antibodies which interact with multiple sites on the patients own

TSH receptor. Structurally, the antibodies appear to have virtually identical properties to antibodies interacting with non-self antigens, with the variable regions of the molecule providing the autoantibody characteristics.

With regard to the autoantigen, all studies to date indicate that the TSH receptor in Graves' disease is identical to that in normal human tissue. These analyses have involved the use of TSH (Smith and Hall, 1974) and TSH-receptor antibodies (Knight and Adams, 1980) as probes and might be expected to be sensitive ways of assessing any change in receptor characteristics which could be involved in initiation of an autoimmune reaction (however, see Section 9.11.4).

Several factors are associated with increased risk in the appearance of TSH-receptor antibodies (development of Graves' disease) in individuals and these include (for review see Friedman and Fialkow, 1978) (i) female sex, (ii) a family history of Graves' disease, (iii) the occurrence of endocrine autoimmune disease in the individual or his family, (iv) the occurrence of certain HLA antigens. Detailed analysis of these observations indicates that the occurrence of Graves' disease is inconsistent with simple autosomal dominant or recessive inheritance. In addition, several examples of apparent male to male transmission of Graves' disease rule out X-linked mechanisms in these cases. Studies of Graves' disease in twins provide a direct means of assessing the relative importance of genetic factors in the pathogenesis of the disease. These observations indicate that about 50 per cent of monozygotic (genetically identical) twin pairs are concordant for hyperthyroidism in contrast to less than 5 per cent of dizygotic (differing on average by half their genes) pairs. The greater rate of concordance for hyperthyroidism (and presumably TSH-receptor antibodies) among monozygotic twins emphasizes the importance of hereditary determinants in the development of Graves' disease. However, non-inherited factors also appear to be involved pathogenically since not all of the monozygotic pairs are concordant.

Consequently, any theory concerning the development of TSH-receptor antibodies must consist of both inherited and non-inherited components and explain the various associations outlined above. In addition, Graves' disease is characterized by onset after many years of life and periods of remission and relapse and these too must be explained by the theory.

To date, two detailed theories have been proposed to explain the appearance of TSH-receptor antibodies and other organ-specific autoantibodies. These are (a) the suppressor T cell defect hypothesis originally proposed by Alison *et al.* (1971) and elaborated by Volpé and co-workers (Kidd *et al.*, 1980) and (b) the V and H gene theories proposed by Adams and Knight (1980; Adams, 1978).

The suppressor T cell theory (reviewed by Kidd *et al.*, 1980) proposes that forbidden clones of cells capable of synthesizing autoantibodies arise by somatic mutation in the normal individual and in the normal individual these clones are immediately suppressed by suppressor T lymphocytes. The theory proposes that organ-specific autoimmune disease results from a specific defect in immuno-regulation by suppressor lymphocytes that permits survival of the appropriate organ-specific self reactive 'forbidden' clone.

In order to explain the HLA association with the development of Graves' disease it has been proposed that the gene responsible for the putative specific defect in immuno-regulation is in linkage disequilibrium with HLA-Dw3 on chromosome 6. Remission and relapse has been attributed to a 'partial defect' in immuno-regulation associated with changes in the patient's 'stress' condition and the association with other endocrine autoimmune disorders by proposing that the genes responsible are closely related.

9.11.2 Evidence for the role of a suppressor T cell defect in Graves' disease and other closely related autoimmune disorders

There is no direct evidence to indicate that a suppressor T cell defect is involved in the development of organ-specific autoantibodies. On the contrary, there is direct evidence to suggest that a suppressor T cell defect is not involved. In particular: (i) T lymphocytes from normal subjects are unable specifically to inhibit thyroid autoantibody synthesis by B lymphocytes from patients with autoimmune thyroid disease (Beall and Kruger, 1979; McLachlan *et al.*, 1980); (ii) the spontaneous development of anti-erythrocyte autoantibodies in NZB mice is not influenced by repeated injections of thymus cells from syngeneic donors (Knight and Adams, 1978).

Indirect evidence for a suppressor T cell defect in autoimmune thyroid disease includes studies with blast transformation, leucocyte migration inhibition, leucocyte adherence to glass and macrophage–lymphocyte rosettes (Kidd *et al.*, 1980). However, the relationship of these 'parameters of cell-mediated immunity' to autoantibody synthesis is not clear at present.

9.11.3 The V-gene and H-gene theories of autoimmunity

As mentioned previously, the autoantibody characteristics of autoantibodies are due to the variable regions of the molecule (Fig. 9.9). The V-gene theory of autoimmune disease (Adams, 1978), can be considered to be developed from the hypothesis that the genes which code for these variable regions are important in the genetic basis of the autoimmune process. It is proposed that in certain combinations these genes form pre-forbidden clones which, in post-natal life are liable to develop into self reactive, forbidden clones through somatic mutation. Once formed, stimulation with autoantigen would be expected to occur.

Recently, Adams and Knight (1980) have been able to make some assessment of the V-gene theory by analysing the genes responsible for the

Variable regions

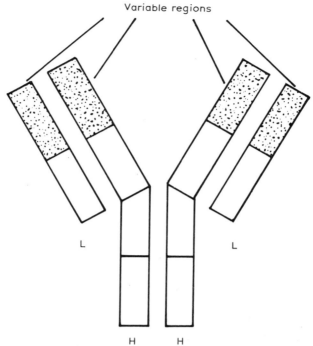

Fig. 9.9 Scheme of IgG structure.

development of lupus nephritis and autoimmune anaemia in New Zealand strains of mice. The studies suggested that histocompatability associated (H)-genes rather than V-genes, were the predominant genetic determinants in the development of these autoimmune diseases. Consequently, they have modified the V-gene theory to accommodate a major role for H-genes (Adams and Knight, 1980).

The H-gene theory postulates that deletion of autoreactive cell clones in foetal life, involves recognition processes dependent on histocompatibility antigens and that the genes (H-genes) coding for these antigens are important in the predisposition to autoimmunity.

It follows from this premise, that the H-genes will be an important factor in determining which clones of immune cells survive the deletion process. This might be expected to be important in at least two ways in the development of autoimmunity. (1) In controlling the survival of pre-forbidden clones which can then develop into autoantibody-synthesizing (forbidden) clones by V-gene somatic mutation. (2) The survival of cell clones which can respond to, and inhibit autoantibody-synthesizing clones, when they appear as a result of somatic mutation (an example of this type of response, is the formation of antibodies to idiotypic determinants on the autoantibody (Jerne, 1973, 1974).

The H-gene theory of autoimmune disease, readily explains the features of Graves' disease which have already been outlined. For example:

  (i) the association with HLA through the histocompatability antigens;

 (ii) the involvement of both inherited and non-inherited factors through processes involving both H-genes and somatic mutation of V-genes;

(iii) the influence of male sex in reducing the incidence of Graves' disease by way of histocompatibility genes on the Y chromosome;

(iv) the development of different but related autoimmune diseases, can be attributed to different repertoires of both pre-forbidden clones and immunoregulatory clones (cells capable of inhibiting autoantibody synthesis;

 (v) spontaneous relapses and remissions in some patients with Graves' disease can be attributed to the ability of the patients' immune system to respond to autoantibody-synthesizing cell clones (by way of Jernes' (1973, 1974) network for example). Variation in the efficiency of this response would be expected to reflect the extent of H-gene-dependent clonal deletion in early life and give rise to different patterns of relapse and remission in different patients as is observed.

The H-gene theory clearly provides a detailed, fundamentally based explanation for the development of organ specific autoimmune diseases such as Graves' disease. Although there is no direct evidence that H-genes are the major genetic determinants in the development of autoimmune disease in man, it should be possible to test many aspects of the hypothesis. In particular, the use of *in vitro* model systems, based on autoantibody synthesis by lymphocytes from patients with autoimmune disease, should permit an assessment of any defect in immunoregulation such as inability to recognize idiotypic determinants on autoantibodies. Also, the prospect of producing monoclonal autoantibodies should make detailed analysis of the final products of the putative H-gene defect possible, as discussed in the following section.

9.11.4 The role of viral infection in the development of autoimmune disease

The role of both inherited and non-inherited factors in the development of Graves' disease, has led to the suggestion that the non-inherited factors could be attributed to a viral infection (Friedman and Fialkow, 1978). Such an effect could conceivably be mediated by way of virally induced minor modifications in the structure of the TSH receptor leading to recognition of the receptor as 'non-self' by the immune system and TSH-receptor antibody formation.

The TSH receptor in Graves' thyroid tissue, appears to have the same TSH and TSH-receptor antibody binding characteristics as normal thyroid tissue and this can be considered to be evidence against a virally induced change in the TSH receptor. However, in studies with experimental autoimmune myasthenia gravis, immunization of mice with acetylcholine receptor prepared from the electric organ of the electric eel, leads to the formation of some antibodies which react with the mouse acetylcholine receptor, but not acetylcholine receptors from the electric eel (Berman and Patrick, 1980). This suggests that the assessment of an unchanged TSH receptor in Graves' disease by hormone and antibody binding studies may be inadequate and when possible, more detailed structural analysis will have to be made to establish if the TSH receptor is modified in Graves' thyroid tissue.

An alternative role for a virus in the development of autoimmune disease is suggested by the V- and H-gene theories. In particular, somatic mutation of pre-forbidden clones could be influenced by viral infection.

9.12 FUTURE DEVELOPMENTS

As the hyperthyroidism of Graves' disease appears to be due to autoantibodies interacting with the TSH receptor and acting as TSH agonists, an increased understanding of the disease process might be expected to arise from a greater knowledge of:

(1) the immunological mechanisms involved in the breakdown in self tolerance, which lead to the formation of organ specific autoantibodies;
(2) the structure of the receptor antibodies;
(3) the nature of the TSH receptor and the mechanisms by which binding of TSH or antibody, can activate adenylate cyclase.

With regard to all three points, the possibility of producing monoclonal autoantibodies by cell fusion, is an exciting prospect for providing further information. Fusion of antibody-synthesizing mouse spleen cells with a mouse myeloma to form a hybridoma secreting monoclonal antibodies, was first described by Kohler and Milstein (1975). It then appeared possible, in principle at least, to extend the fusion system to autoantibody-synthesizing lymphocytes from patients with autoimmune diseases and produce monoclonal human autoantibodies. However, for various technical reasons, creation of hybrid cells capable of synthesizing human antibodies, has not yet been achieved. When the technical problems have been overcome and it is possible to prepare relatively large amounts of monoclonal human autoantibodies, detailed analysis of the structure of these proteins should provide invaluable insight into the nature of the autoimmune process. It should also be possible to use our understanding of the immune system to develop specific forms of therapy for autoimmune diseases such as Graves' disease. In particular, the use of antibodies to the idiotypic determinants on autoantibodies specifically to eliminate autoantibody-synthesizing cell clones, provides an exciting therapeutic prospect.

An understanding of the nature of the TSH receptor and the mechanisms by which receptor binding activates adenylate cyclase requires a detailed knowledge of the structure of the receptor, coupling proteins and cyclase. At present little is known about any of these membrane compounds and the problems of analysing such a complex amphiphilic system, using currently available methodology are considerable. With regard to the ability of some TSH-receptor antibodies to act as TSH agonists, while others act as antagonists however, monoclonal autoantibodies should be useful in providing further information. In particular, it should be possible to isolate monoclonal antibodies with receptor binding-stimulating properties and receptor binding-non-stimulating properties. These preparations should be invaluable probes for analysing the TSH receptor and the mechanism of action of TSH-receptor antibodies.

REFERENCES

Adams, D.D. and Purves, H.D. (1956), *Proc. Univ. Otago Med. Sch.*, **34**, 11–12.

Adams, D.D. and Kennedy, T.H. (1967), *J. Clin. Endocrinol. Metab.*, **27**, 173–177.

Adams, D.D. and Kennedy, T.H. (1971), *J. Clin. Endocrinol. Metab.*, **33**, 47–51.

Adams, D.D., Kennedy, T.H. and Stewart, R.D.H. (1976), *Aust. N.Z.J. Med.*, **6**, 300–304.

Adams, D.D. (1978), *J. Clin. Lab. Immunol.*, **1**, 17–74.

Adams, D.D. and Knight, J.G. (1980), *Lancet*, **i**, 396–398.

Adlkofer, F., Schleusener, H., Uher, L. and Ananos, A. (1973), *Acta Endocrinol.*, **73**, 483–488.

Agnew, W.S. and Raftery, M.A. (1979), *Biochemistry*, **18**, 1912–1919.

Allison, A.C., Denman, A.B. and Barnes, R.D. (1971), *Lancet*, **ii**, 135–137.

Amir, S.M., Sullivan, R.C. and Ingbar, S.H. (1978), *Endocrinology*, **103**, 101–111.

Baxter, J.D. and Funder, J.W. (1979), *N. Engl. J. Med.*, **301**, 1149–1161.

Beall, G.N. and Kruger, S.R. (1979), *J. Clin. Endocrinol. Metab.*, **48**, 712–714.

Benhomou-Glynn, N., El Kabir, D.J., Roitt, I.M. and Doniach, D. (1969), *Immunology*, **16**, 187–204.

Berman, P.W. and Patrick, J. (1980), *J. Exp. Med.*, **151**, 204–223.

Bitensky, L., Alaghband-Zadeh, J. and Chayen, J. (1974), *Clin. Endocrinol.*, **3**, 363–374.

Bolk, J.H., Elte, J.W.F., Bussemaker, J.K., Haak, A. and Van Der Heide, D. (1979), *Lancet*, **ii**, 61–63.

Brennan, A., Rees Smith, B. and Hall, R. (1980), *J. Endocrinol.*, **85**, 43P–44P.

Brown, R.S., Jackson, I.M.D., Pohl, S.L. and Reichlin, S. (1978), *Lancet*, **i**, 904–906.

Clague, R., Mukhtar, E.D., Pyle, G.A., Nutt, J., Clark, F., Scott, M., Evered, D., Rees Smith, B. and Hall, R. (1976), *J. Clin. Endocrinol. Metab.,* **43**, 550–556.

Davies, T.F. (1977), M.D. Thesis, University of Newcastle upon Tyne.

Davies, T.F., Rees Smith, B. and Hall, R. (1978), *Endocrinology,* **103**, 6–10.

Davies, T.F., Yeo, P.P.B., Evered, D.C., Clark, F., Rees Smith, B. and Hall, R. (1977), *Lancet,* **ii**, 1181–1182.

Davoli, C. and Salabè, G.B. (1976), *Clin. Exp. Immunol.,* **23**, 242–247.

Dawes, P.J.D., Peterson, V.B., Rees Smith, B. and Hall, R. (1978), *J. Endocrinol.,* **78**, 89–102.

Dirmikis, S.M., Justice, S.K. and Munro, D.S. (1974), *Biochim. Biophys. Acta,* **379**, 239–246.

Dirmikis, S.M. and Munro, D.S. (1973), *J. Endocrinol.,* **59**, 579–592.

Dirmikis, S.M. and Munro, D.S. (1975), *Br. Med. J.,* **2**, 665–666.

Docteur, R., Bos, G., Visser, T. and Hennemann, G. (1980), *Clin. Endocrinol.,* **12**, 143–153.

Dorrington, K.J. and Munro, D.S. (1966), *Clin. Pharm. Ther.,* **7**, 788–806.

Einhorn, J., Einhorn, N., Fagraeus, A. and Jonsson, J. (1967), in *Thyrotoicosis* (Irvine, W.J., ed.), Livingstone, Edinburgh.

Endo, K., Kasagi, J., Konishi, J., Ikekubo, K., Okuno, T., Takeda, Y., Mori, T. and Torizuka, K. (1978), *J. Clin. Endocrinol. Metab.,* **46**, 734–739.

Fenzi, G.F., Hashizume, K., Roudebush, C.P. and DeGroot, L.J. (1979), *J. Clin. Endocrinol. Metab.,* **48**, 572–578.

Friedman, J.M. and Fialkow, P.J. (1978), *Clin. Endocrinol. Metab.,* **7**, 47–65.

Gill, D.L., Marshall, N.J. and Ekins, R.P. (1977), *Biochem. Soc. Trans.,* **5**, 1064–1067.

Ginsberg, J., Rees Smith, B. and Hall, R. (1980), *J. Endocrinol.,* **87**, 33P–34P.

Hales, I.B., Lutterell, B.M. and Saunders, D.M. (1980), *Proc. VIII Int. Thyroid Congr.,* Sydney, Australia, 591–593.

Hall, R. and Rees Smith, B. (1979), *Clinical Immunology Update* (Fanklin, E.C., ed), Elsevier, New York, pp. 291–304.

Hall, R., Rees Smith, B. and Mukhtar, E.D. (1975), *Clin. Endocrinol.,* **4**, 213–230.

Hart, I. and McKenzie, J.M. (1971), *Endocrinology,* **88**, 26–30.

Hashizume, K., Fenzi, G. and De Groot, L.J. (1978), *J. Clin. Endocrinol. Metab.,* **46**, 679–689.

Jerne, N.K. (1973), *Sci. Am.,* **229**, 52–60.

Jerne, N.K. (1974), *Ann. Immunol.* (Inst. Pasteur), **125c**, 373–389.

Kendall-Taylor, P. (1972), *J. Endocrinol.,* **52**, 533–540.

Kendall-Taylor, P. (1973), *Br. Med. J.,* **3**, 72–75.

Kendall-Taylor, P. and Munro, D.S. (1971), *Biochim. Biophys. Acta,* **231**, 314–319.

Kidd, A., Okita, N., Row, V.V. and Volpe, R. (1980), *Metabolism,* **29**, 80–99.

Knight, J.G. and Adams, D.D. (1978), *J. Clin. Lab. Immunol.,* **1**, 151–158.

Knight, A. and Adams, D.D. (1980), *Hormone Res.,* **13**, 69–80.

Kohler, G. and Milstein, C. (1975), *Nature,* **256**, 495–497.

Konishi, J., Kasagi, K. Endo, K., Mori, T., Torizuka, K., Yamada, Y., Nohara, Y., Matsuura, N. and Kojima, H. (1980), *Proc. VIII Int. Thyroid. Congr.,* Sydney, Australia, 555–558.

Kriss, J.P. (1968), *J. Clin. Endocrinol. Metab.,* **28**, 1440–1444.

Kuzuya, N., Chiu, S.C., Ikeda, H., Uchimura, H., Ito, K. and Nagataki, S. (1979), *J. Clin. Endocrinol.*, **48**, 706–711.

Lonergan, D., Babiarz, D. and Burke, G. (1973), *J. Clin. Endocrinol. Metab.*, **36**, 439–444.

Maisey, M.N. (1972), *Clin. Endocrinol.*, **1**, 189–198.

Manley, S.W., Bourke, J.R. and Hawker, R.W. (1974), *J. Endocrinol.*, **61**, 437–445.

McGregor, A.M., McLachlan, S.M., Rees Smith, B. and Hall, R. (1980a), *J. Mol. Med.*, **4**, 119–127.

McGregor, A.M., McLachlan, S.M., Rees Smith, B. and Hall, R. (1979b), *Lancet*, **ii**, 442–444.

McGregor, A.M., McLachlan, S.M., Rees Smith, B. and Hall, R. (1980b), *N. Engl. J. Med.*, **303**, 302–307.

McGregor, A.M., Peterson, M.M., Capifferi, R., Evered, D.C., Rees Smith, B. and Hall, R. (1978), *J. Endocrinol.*, **81**, 114P–115P.

McGregor, A.M., Peterson, M.M., Capiferri, R., Evered, D.C., Rees Smith, B. and Hall, R. (1979a), *Clin. Endocrinol.*, **11**, 437–444.

McGregor, A.M., Rees Smith, B., Hall, R., Petersen, M.M., Miller, M. and Dewar, P.J. (1980c), *Lancet*, **i**, 1101–1103.

McLachlan, S.M., McGregor, A.M., Rees Smith, B. and Hall, R. (1979), *Lancet*, **i**, 162–163.

McLachlan, S.M., McGregor, A.M., Rees Smith, B. and Hall, R. (1980), *J. Clin. Lab. Immunol.*, **3**, 15–21.

McKenzie, J.M. (1958a), *Endocrinology*, **62**, 865–868.

McKenzie, J.M. (1958b), *Endocrinology*, **63**, 372–382.

McKenzie, J.M. and Zakarija, M. (1976), *J. Clin. Endocrinol. Metab.*, **42**, 778–781.

Mehdi, S.Q. and Nussey, S.S. (1975), *Biochem. J.*, **145**, 105–111.

Mukhtar, E.D., Rees Smith, B., Pyle, G.A., Hall, R. and Vice, P. (1975), *Lancet*, **i**, 713–715.

Munro, D.S., Dirmikis, S.M., Humphries, H., Smith, T. and Broadhead, G.D. (1978), *Br. J. Obstets. Gyn.*, **85**, 837–843.

Nutt, J., Clark, F., Welch, R.G. and Hall, R. (1974), *Br. Med. J.*, **iv**, 695–696.

O'Donnell, J., Trokoudes, K., Silverberg, J., Row, V. and Volpé, R. (1978), *J. Clin. Endocrinol. Metab.*, **46**, 770–777.

Ochi, Y. and De Groot, L.J. (1968), *Endocrinology*, **83**, 845–854.

Ochi, Y., Yoshimura, M., Hachiya, T. and Miyazaki, T. (1977), *Acta Endocrinol.*, **85**, 791–798.

Onaya, T., Kotani, M., Yamada, T. and Ochi, Y. (1973), *J. Clin. Endocrinol. Metab.*, **36**, 859–866.

Ozawa, Y., Maciel, R.M.B., Chopra, I.J., Solomon, D.H. and Beall, G.N. (1979), *J. Clin. Endocrinol. Metab.*, **48**, 381–387.

Petersen, V.B., Dawes, P.J.D., Rees Smith, B. and Hall, R. (1977), *FEBS Lett.*, **83**, 63–67.

Petersen, V.B., Rees Smith, B. and Hall, R. (1975a), *J. Clin. Endocrinol. Metab.*, **41**, 199–202.

Petersen, V.B., Smith, B.R. and Hall, R. (1975b), *Thyroid Research* 7th *Int. Thyroid Conf.*, Boston, pp. 610–613.

Pinchera, A., Liberti, P., Martino, E., Fenzi, G.F., Grasso, L., Rovis, L., Baschieri, L. and Doria, G. (1969), *J. Clin. Endocrinol. Metab.*, **29**, 231–238.

Rees Smith, B. and Hall, R. (1980), *Proc.* VIII *Int. Thyroid Congr.*, Sydney, Australia, 715–716.

Rees Smith, B., Mukhtar, E.D, Pyle, G.A., Kendall-Taylor, P. and Hall, R. (1975), *Thyroid Research. Proc.* 7th *Int. Thyroid Congr.*, Boston. *Excerpta Medica*, pp. 411–413.

Rees Smith, B., Pyle, G.A., Peterson, V.B. and Hall, R. (1977a), *J. Endocrinol.*, **75**, 391–400.

Rees Smith, B., Pyle, G.A., Peterson, V.B. and Hall, R. (1977b), *J. Endocrinol.*, **75**, 401–407.

Schleusener, H., Kotulla, P., Finke, R., Sorje, H., Meinhold, H., Adlkofer, F. and Wenzel, K.W. (1978). *J. Clin. Endocrinol. Metab.*, **47**, 379–384.

Shuman, S.J., Zor, U., Chayoth, R. and Field, J.B. (1976), *J. Clin.* Invest **57**, 1132–1141.

Smith, B.R. (1970), *J. Endocrinol.*, **46**, 45–54.

Smith, B.R. (1971), *Biochem. Biophys. Acta*, **229**, 649–662.

Smith, B.R., Dorrington, K.J. and Munro, D.S. (1969a), *Biochim. Biophys. Acta*, **192**, 277–285.

Smith, B.R., Munro, D.S. and Dorrington, K.J. (1969b), *Biochim. Biophys. Acta*, **188**, 89–100.

Smith, B.R. and Hall, R. (1974), *Lancet,* **ii**, 427–431.

Smith, B.R. and Munro, D.S. (1970), *Biochim. Biophys. Acta,* **208**, 285–293.

Söderström, N. and Björklund, A. (1974), *Scand. J. Immunol.,* **3**, 295–302.

Solomon, D.H., Chopra, I.J., Chopra, U. and Smith, F.J. (1977), *N. Engl. J. Med.*, **296**, 181–186.

Sugenoya, A., Kidd, A., Row, V.V. and Volpé, R. (1979), *J. Clin. Endocrinol. Metab.*, **48**, 398–402.

Swanson Beck, J., Young, R.J., Simpson, J.G., Gray, E.S., Nicol, A.G., Pegg, C.A.S. and Michie, W. (1973), *Br. J. Surg.,* **60**, 769–771.

Teng, C.S., Rees Smith, B., Anderson, J. and Hall, R. (1975), *Biochem. Biophys. Res. Commun.,* **66**, 836–841.

Teng, C.S., Rees Smith, B., Clayton, B., Evered, D.C., Clark, F. and Hall, R. (1977), *Clin. Endocrinol.,* **6**, 207–211.

Teng, C.S. and Yeung, R. (1980), *J. Clin. Endocrinol. Metab.*, **50**, 144–147.

Van Herle, A.J., Vassart, G. and Dumont, J.H. (1979), *N. Engl. J. Med.,* **301**, 307–315.

Verrier, B., Fayet, G. and Lissitzky, S. (1974), *Eur. J. Biochem.,* **42**, 355–365.

Wong, E.T. and Doe, R.P. (1972), *Ann. Intern. Med.,* **76**, 77–84.

Zakarija, M. and McKenzie, J.M. (1978a), *Endocrinology,* **103**, 1469–1475.

Zakarija, M. and McKenzie, J.M. (1978b), *J. Clin. Endocrinol. Metab.,* **47**, 249–254.

Zakarija, M. and McKenzie, J.M. (1978c), *J. Clin. Endocrinol. Metab.,* **47**, 906–908.

Index